California Catastrophes

MW00576067

California Catastrophes

The Natural Disaster History
of the Golden State

Gary Griggs

UNIVERSITY OF CALIFORNIA PRESS

University of California Press
Oakland, California

© 2024 by Gary Griggs

Library of Congress Cataloging-in-Publication Data

Names: Griggs, Gary B., author.
Title: California catastrophes : the natural disaster
   history of the Golden State/ Gary Griggs.
Description: Oakland, California : University of
   California Press, [2024] | Includes bibliographical
   references and index.
Identifiers: LCCN 2023056545 (print) | LCCN 2023056546
   (ebook) | ISBN 9780520402089 (cloth) | ISBN
   9780520402102 (paperback) | ISBN 9780520402119
   (ebook)
Subjects: LCSH: Natural disasters—California—History.
Classification: LCC GB5010 .G75 2024 (print) |
   LCC GB5010 (ebook) | DDC 363.3409794—dc23/
   eng/20240131
LC record available at https://lccn.loc.gov/2023056545
LC ebook record available at https://lccn.loc
   .gov/2023056546

Manufactured in the United States of America

33 32 31 30 29 28 27 26 25 24
10 9 8 7 6 5 4 3 2 1

Contents

Figures and Tables

x | List of Figures and Tables

TABLES

Acknowledgments

No book is entirely the product of a single individual. In particular, I want to acknowledge and thank my wife and partner, Deepika Shrestha Ross, for working alongside me on the multiple details necessary to bring any writing project through what sometimes seems like a long, dim tunnel to the light at the other end. She not only is a creative editor, but was exceptionally talented and patient in tracking down every photograph, map, and graphic; obtaining the necessary permissions; and then acknowledging each individual. This book would not have happened without her.

Two things have continued to encourage my writing and teaching over the years of studying natural disasters in California. The first is that these events keep coming. Not too many months pass before there is another earthquake or landslide, flood or drought, El Niño or wildland fire. Related to this sense of geological anxiety are both the questions and interest about these natural disasters that come from my classes, public lectures, and writing, which encouraged me to undertake this project. I am grateful for the student interest in my classes and the general public whose interest in natural disasters keeps me inspired.

No writer is an author without a publisher, and I am indebted to the staff at the University of California Press, specifically Chloe Layman and Chad Attenborough, for seeing the value of this book, taking on this project, and guiding me through the multiple steps needed to go from an early manuscript to the final book.

Introduction to California's Natural Disasters

There is science, logic, reason; there is thought verified by experience. And then there is California.

—Edward Abbey, *A Voice Crying in the Wilderness*

I'm a Californian, who has never been tempted to settle down anywhere else in the world (though traveling is always a joy), and still, Edward Abbey's words resonate with me. California is a bit of a conundrum, a paradox, a paradise that includes more than its share of perils. While the state's reputation may have lost some of its luster in recent years, representations of California in popular culture—through literature, music, film, fashion, food, and sports—that now reach almost every corner of the world have firmly established its mystique as a place of fairy tales, opportunity, and possibly reinvention.

Fittingly, the origin of the name *California*, a subject of some debate, is now generally agreed to have come from an early-16th-century Spanish romance novel, *Las Sergas de Esplandián* (*The Adventures of Esplandián*). Centuries after its publication, historian Edward Everett Hale (the grandnephew of Nathan Hale, American military officer and spy during the Revolutionary War) came across the long-forgotten book in which the name of a mythical island populated only by black warrior women is—you guessed it—California. Armed with the book and some additional evidence, Hale made the case that when the early Spanish explorers came upon the Baja California peninsula, they named it California after the fictional island in the book, because they hadn't explored far enough north into the Gulf of California to know that it was actually a peninsula and not an island.

This is an excerpt from the March 1864 issue of *Atlantic Monthly*, which published an English translation of the novel:

> Know then that on the right hand of the Indies, there is an island called **California**, very close to the side of the Terrestrial Paradise, and it was peopled by black women, without any man among them, for they lived in the fashion of Amazons. They were of strong and hardy bodies, of ardent courage and great force. Their island was the strongest in all the world, with its steep cliffs and rocky shores. Their arms were all of gold, and so was the harness of the wild beasts which they tamed and rode. For, in the whole island, there was no metal but gold.
>
> —*Las Sergas de Esplandián*, by Garci Rodríguez de Montalvo—Seville, Spain 1510

As a state, California is full of superlatives and extremes. It has the highest and lowest points in the lower 48 states (14,805 feet at Mount Whitney and −282 feet in Death Valley). The right combination of soil and sunlight—and water, historically—allowed California to become a cornucopia of crops. Today the state grows over a third of the nation's vegetables and two-thirds of the country's fruits and nuts, including 100 percent of the almonds, 99 percent of the walnuts and artichokes, 97 percent of the kiwis and plums, 95 percent of the celery and garlic, 90 percent of the Brussels sprouts, 89 percent of the cauliflower, 71 percent of the spinach, and 69 percent of the carrots (and this list actually goes on much further).

The computer age had its birth in Northern California, and Southern California arguably remains the epicenter of film and music. Were California a sovereign nation, its economy would now rank fifth largest in the world, only surpassed by the United States, China, Japan, and Germany and ahead of the United Kingdom, India, France, and all of the other 187 countries on the planet. This is no small accomplishment.

California can also claim fame for having more natural hazards per square mile than any other state in the nation. This fact has not, however, deterred people from moving here, as California also happens to be the most populous state in the country, with the largest percentage of immigrants as well.

Going back in time, the earliest residents were the Native Americans who arrived from the north after crossing the Bering land bridge from Asia during the last ice age. Thousands of years later, the Spanish who had established a foothold in Baja California pushed northward. Their Franciscan priests established the mission system that instigated the destruction of the Native American way of life for tribes along the coast, decimating their population through the introduction of new diseases. During the Mexican era, after the missions were dismantled, numerous land grants brought settlers from Mexico northward. Later, settlers of European origin came across the plains in covered wagons from the east. Additional immigrants also arrived from Asia across the Pacific Ocean, contributing to development of the state's infrastructure and economy despite challenging conditions and exclusionary legislation.

People have flocked to California for 175 years with no letup, at least until the last two years—why? What drew people here, whether from the Midwest, New England, or elsewhere? Didn't they know we have earthquakes here—really big ones on occasion? And how about fires, droughts, and floods? The gold rush that started in 1849 cemented the state as a land of promise and opportunity, where just about anything seemed possible. Certainly, the moderating climate has been a dominating draw; after freezing winters across the Midwest and East Coast, along with the lack of fresh fruits and vegetables, California from afar can seem like paradise.

Economic opportunities and jobs have always been attractions as well, beginning on a large scale with the Dust Bowl, with its severe drought and dust storms in the Great Plains states in the 1930s. The need for farmworkers has continued to expand and brought immigrants in from the state's southern border. The expansion and suburbanization of greater Los Angeles and the San Francisco Bay and Santa Clara Valley areas brought more immigrants and job opportunities in the aircraft and then the aerospace industries during and after the World War II years. These years transitioned into the tech age, which had its birth in Silicon Valley in 1939 when William Hewlett and David Packard first built electronic test and measurement equipment in a small garage in Palo Alto. And the growth and expansion of the state's economy continued and now make it economically the fifth-largest nation on the planet, and about to surpass number four, Germany.

New arrivals, drawn by the climate, the landscape, and opportunity, have added to California's richness and diversity. Novel ideas, new patents, breakthrough discoveries, and innovative developments have continued to keep California in the forefront. But not far beneath the surface, tectonic forces continue to grind away, building up to the next "big one." In the late 1960s, BBC produced a film entitled *San Francisco—The City That Waits to Die* that shook things up for a while and still does on occasion. And Hollywood continues to exploit geologic disasters for fun and profit: *San Andreas, Earthquake,* and *Volcano* are just a few disaster films where geologists become the protagonists. These movies have perhaps lulled viewers into the belief or perception that these disasters are just Hollywood constructs, much like its theme parks, and therefore aren't hazards they need to take seriously. There are just too many other serious things in our daily lives to worry about earthquakes or volcanoes.

Longtime San Francisco resident and writer, the late Curt Gentry, wrote (among many other books) *The Last Days of the Late, Great State of California* in 1968, which captured the culture and politics of the state under Governor Ronald Reagan. The book opens with a somewhat sensational and catastrophic account of the San Andreas Fault unzipping from Point Arena in the north to the Salton Sea in the south, with all of California west of the fault, along with its 15 million people, plunging off into the Pacific Ocean during a massive earthquake. While this colossal event itself is fictional, the rest of the book is a nonfiction account of California in the late 1960s and how events in the state created revolutions well beyond the state's borders. In response to the book's publication and digestion, some fringe religious believers in the Los Angeles area actually decided the end was coming and moved away.

Unfortunately, earthquakes are just one of a myriad of natural hazards that wreak havoc on California and its residents. There are also floods, droughts and fires, landslides and mudflows, and coastal storms (their frequency and intensity further exacerbated by climate change).

The earliest explorers in California quickly became aware of the diverse landscape and topography of the state and the difficulties of traversing the Sierra Nevada and the Coast Ranges, particularly after the relatively easy travel across the monotonous plains and prairies of the midcontinent. Interstate and state highways have eliminated some of the adventure and much of the danger, but the state's formidable topography is still hard to avoid, and winter road closures are still common (figure 1.1).

However, it wasn't until the great San Francisco earthquake (and fire) of 1906 that California residents first became aware of the state's dynamic nature and geological unrest. To the surprise of nearly everyone living here at the time, a very large fragment of coastal California—hundreds of miles long—slid northwest as much as 20 feet along the San Andreas Fault, leading to a near-complete loss of the city of San Francisco and heavy damage to many other nearby cities as well

FIGURE 1.1. Failure at Rat Creek along Highway One (Big Sur), winter 2021. *Courtesy of Caltrans, public domain.*

(figure 1.2). Subsequent major earthquakes occurred in Santa Barbara in 1927, Long Beach in 1933, Bakersfield in 1952, the San Fernando Valley in 1971 and again in 1994, and the Santa Cruz Mountains in 1989. All are decadal reminders that California is in its geological adolescence and its terrestrial processes continue to play out, never lying dormant for long.

Another glimpse of the area's unstable geological foundation came a little over a century ago in 1915 when Lassen Peak, one of the southernmost of the chain of Cascade volcanoes that stretches from Northern California into British Columbia, erupted (figure 1.3).

The striking features of California's diverse landscape—the Sierra Nevada and the Cascade volcanoes, the San Andreas Fault and its associated earthquakes, the rugged coastal mountains, and the steep coastal cliffs and uplifted marine terraces that have made much of the state's coastal development possible—all have their origins in millions of years of large-scale tectonic forces that continue today.

Having studied, written, and lectured about geologic hazards for many years, I am often asked by friends why I love disasters. I am not really enamored with natural disasters, but I do feel a sense of wonder at the beauty and power of physical processes, which quickly lead to questions of how, and why, and when? Like many traditional geologists, I have spent much of my early professional life looking at outcrops—mountainsides, roadcuts, river canyons, or sea cliffs—acting

FIGURE 1.2. Ruins of San Francisco in the vicinity of Post and Grant Avenues following the 1906 San Francisco earthquake and fire. *By H. D. Chadwick, public domain, via Wikimedia Commons.*

FIGURE 1.3. The 1915 eruption of Lassen Peak. *By B. F. Loomis, courtesy of US Geological Survey, public domain.*

as a forensic detective, deciphering what the exposure is telling me about what happened here 10 or 100 million years ago. By contrast, in studying modern geologic hazards, I can observe firsthand exactly what happens during a real event—a large debris flow, for example—and what kind of record might be preserved from the past.

While the diverse topography of the state—the mountains and valleys, river canyons and floodplains, uplifted marine terraces, and sea cliffs—owes its relief and origins to large-scale tectonic events, there are also numerous surface processes, such as wave attack, intense or prolonged rainfall and runoff, and landslides or other mass downslope movements of rock and soil (and occasionally homes and highways), that continue to alter these features. In addition, sea level along the coast has changed continuously through time such that the position of the shoreline, which coastal homeowners think of as fixed and permanent, is only a temporary one. Although these changes are not rapid, it is clear from geological evidence, historic photographs, and our own observations that the entire state of California, as a result of its geological youth, is alive and active and will continue to be a work in progress.

The geological and climatic processes discussed in the following chapters only become hazards and present risks when people expecting a degree of "permanence" are added to the mix. There is every reason to believe that the geologic and hydrologic processes that have shaped the state's landforms in the past will continue far into the future. Whether earthquakes or landslides, droughts or floods, the only certainty in California is that change is inevitable, and the landscape and climate, both altered by human activity and intervention, are subject to change without much notice.

Chapters 2, 3, and 4 are focused on the large-scale hazards resulting from the tectonic setting of California. These hazards are shared by our neighbors to the north, Oregon and Washington, as well as most of the nations around the rim of the Pacific Ocean. Earthquakes where massive tectonic plates collide or slide alongside each other (figure 1.4), along with the tsunamis that are typically generated by these very large earthquakes, as well as the associated volcanic eruptions, are explained and illustrated—why these events occur where they do, how frequently they take place, and what can happen when the Earth moves. These are all widespread global processes, and California just happens to be sitting astride a boundary between two moving plates—the San Andreas Fault—and we are profoundly affected by the periodic but unpredictable

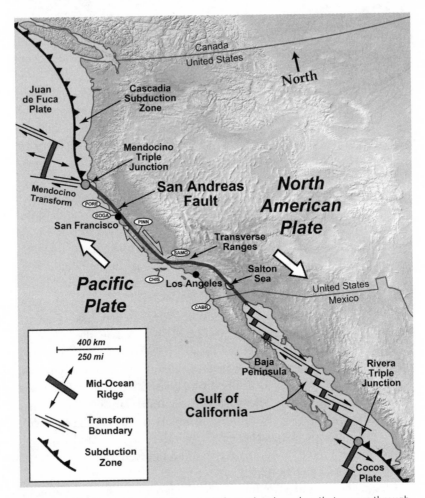

FIGURE 1.4. The San Andreas Fault is a transform plate boundary that passes through western California. *Courtesy of National Park Service, illustration modified from* Earth: Portrait of a Planet, *by S. Marshak, 2001, W. W. Norton & Company, New York.*

movement along that boundary. We are a huge natural laboratory of geologic unrest, and we can expect this to continue for millions of years into the future. Which parts of the state are most vulnerable to these hazards and how might we prepare for these inevitable and potentially catastrophic events and reduce their toll on human lives and our towns and cities?

Chapters 5, 6, and 7 cover processes and hazards resulting from our weather and climate—our hydrometeorological setting. Climate has been described as what we predict, and weather as what we get. Climate is long term, years to centuries; weather is short term, days to weeks. Global climate is changing, however, and California is one of many areas that is experiencing the impacts of those changes. Global climate models and our personal observations are in agreement—summers are getting hotter and drier with prolonged droughts and more frequent and expansive wildland fires with greater property losses, so much so that two major insurance companies in 2023 announced that they were no longer going to insure homes in California. This is serious business with major economic impacts. A warming offshore ocean leads to greater evaporation and additional atmospheric moisture, which is concentrating precipitation in the winter months and generating more frequent and severe flooding, landslides, and debris flows. And as if to make this point more evident, the winter of 2023 saw record precipitation bringing devastating flooding and widespread damage in California. Climate and weather are beyond our control.

Chapter 8 covers mass downslope movements of material, whether rock, soil, or a combination of the two, which can fall, slide, or flow downhill and create death and destruction in its path. These processes are widespread in California and range from large rocks falling onto roadways, to slow-moving landslides that may involve homes, roads, and utilities, and also mud or debris flows that, with little warning, can travel faster than any of us can run. Many of these failures involve excess water from intense or prolonged rainfall, as well as potentially unstable slopes.

In chapter 9, both short- and long-term coastal hazards are illuminated. While sea level is rising at an accelerated rate as the climate continues to warm, due to glacial melt and the expansion of a warmer ocean, over the next several decades it will be the short-term events that will be most damaging. These include strong El Niño events and the arrival of large storm waves coincident with very high tides, as the California coast experienced in January of 1983 and exactly 40 years later in January of 2023 and 11 months later in December of 2023. As with the processes covered in earlier chapters, these are not events we have any control over. The hazards are simply due to where we have chosen to build homes, businesses, and infrastructure. The distance to the shoreline or how close to the edge of a coastal bluff those structures are is what will determine the future fate of these buildings. And some of California's most coveted real estate, whether Malibu, Newport Beach, or Del Mar, has the dubious distinction of also being the most vulnerable to future sea-level rise.

Chapter 10—Where Do We Go from Here?—is a bold effort to bring all of these hazards and risks together, with a discussion of where we are in California today, why people continue to move here, and what the future holds, from a personal perspective. What are the lessons we have learned from past geologic disasters, or what should we have learned? Have we "designed with nature," as the Scottish planner Ian McHarg recommended over half a century ago? Can we reduce further losses through a variety of zoning ordinances and building codes, or are all of our vulnerable communities really at the mercy of future natural hazards, regardless of our best intentions?

With the exception of five years living in Oregon, a year on a Fulbright fellowship in Greece, 6 months conducting research in New Zealand, and 12 months on and off ships at sea, my entire life has been spent in California, and I have admittedly been strongly influenced by my geographic and geologic surroundings. My seven decades in California have been frequently interrupted by earthquakes, droughts and fires, floods and debris flows, catastrophic coastal storms and shoreline retreat, sometimes virtually in my backyard, but elsewhere around the state as well. These geologic hazards have shaped California, contributing to some of its unique physical characteristics. And despite these recurring hazards, like many other residents (whether here by accident or intention), I have decided the rewards of living in California (real or imagined) are worth the risks. In each chapter I have included my own personal experience with each of these hazards, to provide some perspective on living with the inevitable risks we face in the state.

It is for California's residents (and others who may be headed this way or aren't fully aware of what their chances are of suffering through a major geologic disaster) that I write this book. We are here and we plan to stay. We'd like our children and grandchildren to flourish here as well. But the hazards are real, and they are plentiful. The chapters that follow explore each of these hazards and how they have impacted the state and its people over the past 150 or so years. Why do these events occur here? How has climate change affected them? And what might we expect in the future? These are followed by a final chapter providing some overall perspective on the risks we all face in our daily lives and where we might go from here. It is my hope that fortified with this knowledge, we will be more fully equipped to collectively ask thoughtful questions and make wise and equitable choices that consider the needs of all Californians, and not just those with access to power and money and a microphone.

2

Earthquakes and Faulting

At about 5:00 p.m. on October 17, 1989, I was driving south on State Route One along the southern Monterey Bay coastline heading towards Carmel Valley, where I was going to meet up with my mother and brother. The car was bouncing around a bit from the wind, which is common in this area of coastal sand dunes. I happened to be listening to the radio as I noticed a few cars pulled over on the side of the highway and then heard the DJ who was reporting from Salinas say, "We're having an earth . . ." and the radio went dead . . . complete silence, nothing. I also noticed there was some sand sliding down the inland faces of some of the dunes alongside the highway, as the car continued to shake around for a while longer.

I didn't think much more about it until I arrived at my brother's home in Carmel Valley 30 minutes later, where he quickly informed me—knowing I had more than a casual interest in earthquakes—that there had been a major earthquake and one span of the Bay Bridge had collapsed. The first game of the Bay World Series between the San Francisco Giants and the Oakland A's was just about to start when the stadium lights began to sway, beginning a period of complete chaos around the bay that was to last for days.

On my drive home in the dark later than evening, I noticed almost all lights along Highway One were out. When I got close to Watsonville, traffic had been rerouted off of the highway because a bridge over the

Watsonville Slough had collapsed. Driving through the city of Watsonville, I could see fires burning in vacant lots and fields where residents were trying to keep warm, having moved outside their shaken homes. I have to say it all appeared quite ominous.

The only visible lights I could see, driving into Santa Cruz as I got closer to home, were from large searchlights downtown, which I was later to learn were being used to search for survivors from several older masonry buildings that had collapsed. When the impacts of the disaster had been uncovered and documented, the 6.9 magnitude earthquake had been responsible for 63 deaths and 3,757 injuries. Forty-two of the deaths were from the collapse of the top level of a double-decker stretch of freeway along the waterfront of Oakland that had been built on weak mud (figure 2.1). Direct damage from the earthquake was estimated at $6.9 billion, with 18,306 houses damaged and 963 destroyed. Being part of the team evaluating the impacts of the earthquake was to consume a lot of my days and hours over the next 12 months.

While as an almost lifelong resident of California I had experienced a number of small shocks, the 1989 Loma Prieta earthquake was a beast on a much larger scale, with some major impacts that took two decades to repair. This event also raised the question of how much can

FIGURE 2.1. Collapse of the upper deck of the Cypress Freeway in Oakland during the 1989 Loma Prieta earthquake led to 42 fatalities. *Courtesy of US Geological Survey, public domain.*

we do or afford to do in order to plan or prepare for potentially disastrous events that only occur very infrequently but may be devastating and deadly when they do.

GLOBAL EARTHQUAKES

The first initial rumblings of the most destructive earthquake in over 400 years were heard on a stormy morning in Beijing, China, in July 1976. Two earthquakes, the first of magnitude 8 and a second 16 hours later of magnitude 7, tore through one of China's most densely populated regions, crumpling dams and toppling buildings. When it was over, the industrial city of Tangshan was essentially destroyed, and over 250,000 people had died. The high population density; the typical unreinforced mud-brick, tile-roofed buildings; and the fact that the first quake struck when most people were indoors sleeping all contributed to the enormity of the disaster.[1]

The 1976 Tangshan earthquake was recent history when, from the perspective of earthquake fatalities, the 21st century began ominously, and the death and devastation unfortunately continued on what seemed like a regular basis. A magnitude 7.6 earthquake in India led to over 20,000 deaths in 2001. Two years later, Iran suffered a magnitude 6.6 event that claimed over 26,000 lives. The next year, 2004, the enormous 9.1 earthquake offshore in the Sumatra Trench generated the greatest fault rupture of any earthquake ever recorded, spanning a distance along the trench of 900 miles, or longer than the entire state of California. Severe shaking was sustained for about 10 minutes, and the combined fatalities from the earthquake and tsunami reached about 228,000 around the northern Indian Ocean.

Just a year later in 2005, over 87,000 died during a 7.6 magnitude shock in Kashmir, in the Himalayas, and then a 7.9 magnitude earthquake in Wenchuan, China, led to over 87,000 more fatalities in 2008. A large earthquake (magnitude 7.0) struck Haiti two years later, creating near-total destruction in unreinforced-masonry buildings. There was never complete agreement on the death toll, which ranged from a low of about 50,000 from a US government report to over 300,000 initially reported by the government of Haiti—some felt in an effort to obtain more international aid. The most common number of fatalities cited is 230,000, but by any measure it was a disaster for Haiti. In striking contrast, and as a testimony to differences in building codes, construction methods, and materials, the 1989 Loma Prieta earthquake in the Santa Cruz Mountains of essentially the same magnitude (6.9) took 63 lives; a tragedy nonetheless, but on a vastly different scale.

On March 11, 2011, the planet experienced the second massive earthquake of the 21st century, one of the largest ever recorded, offshore of Japan when the Pacific Plate broke loose and slid westward beneath the Eurasian Plate, producing a 9.0 magnitude shock followed by a devastating tsunami. As of July 2023, the number of confirmed deaths was 19,759 according to the reconstruction agency, although even years after the quake, more than 2,500 people were still reported missing. And in April of 2015, a 7.8 magnitude earthquake struck Nepal, where the Indian Plate pushes into the Eurasian Plate, with the destruction of many ancient buildings in Kathmandu and the loss of nearly 9,000 lives.

And these were only the most devastating of hundreds of other moderate to large earthquakes around the world in the first two decades of the 21st century. But these eight major events alone led to the combined loss of over 707,000 lives, or an average of 32,000 each year of the new century.[2]

Most of us in California probably think of the 7.8 magnitude 1906 San Francisco earthquake, in which over 3,000 people died, as a major disaster, and it was for California. Although most of San Francisco was destroyed, due in large part to the fires that followed the earthquake (see figure 1.2), the number of casualties was relatively low for such a large shock. In 1923, an earthquake of similar magnitude near Tokyo killed 143,000 people.

Although California, because of its active and adolescent geology, contains more natural and diverse landscape wonders than most areas of similar size anywhere in the world, it is perhaps best known to those who don't live here for just one thing—earthquakes. California is indeed earthquake country, and the accounts and images of past shocks have left permanent impressions with people across the country and around the world, yet people have continued to move to California for 175 years.

Even if the accounts and events accompanying past earthquakes have been exaggerated, California's position in the seismic world, like the economic world, is an important one. The state lies on a belt of active faults that circles the entire Pacific Ocean, and which is responsible for about 80 percent of the world's earthquakes every year. California's residents experience thousands of earthquakes annually, of which only about 500 are large enough to be felt somewhere. Every four or five years a large, potentially destructive earthquake strikes California, with the greater San Francisco Bay area experiencing about 12 damaging shocks per century, at least during the last 200 years or so of written history. We don't really have significantly more earthquakes than other similar-size areas around the Pacific Rim, but with California being the nation's most populous state with about 39 million people, being paradise or the Garden of Eden in the minds of many who don't live here, and being home to so many immigrants, earthquakes here often seem to get more attention across the country and around the world than those in many other places.

Before giving up all hope, however, remember that all earthquakes are not created equal. The small and moderate-size events that rattle our windows, doors, and chandeliers are relatively common and painless but serve as regular reminders that we live in earthquake country and that the large tectonic plates we live on are constantly shifting and adjusting. These minor disturbances give us pause to think about what we will do when the "big one" takes place, and then most of us calmly forget about it and go back to our computer screens, iPads, or iPhones. We nearly always seem to have more pressing or immediate things to worry about.

Unfortunately there won't be a single big one in California; there have been many big ones and there will be many more. Not frequently, but they will continue to occur. They have to, for one very simple reason—California straddles the boundary between the North American Plate to the east and the Pacific Plate to the west (see figure 1.4). In many of the world's earthquake-prone regions, the plate boundaries, fault lines, and fractures are largely invisible, many lying deep beneath the ocean surface. In California, however, the boundary between these massive plates is called the San Andreas Fault, and it is visible from the air and ground in many places (figure 2.2). It is a ragged scar, like a zipper, crossing the landscape for 700 miles from the Gulf of California to Cape Mendocino, north of San Francisco, where it dives off into the Pacific. The San Andreas and its branches slice through farms and forests, through cities, under reservoirs, freeways, and universities. Over 20 million people live near this massive break in the Earth's crust and its associated fractures. Their lives and property are all affected by the fault and will continue to be for centuries to come.

About 40 miles southeast of San Jose, a major offshoot of the San Andreas, the Calaveras Fault, is slowly tearing the city of Hollister apart. It's well worth the drive to this tidy little farm town on a Saturday or Sunday when everyone else is driving to the beach in Santa Cruz. You don't have to be a geologist to appreciate that the streets, curbs, fences, and houses are gradually being pulled apart as the plates on

FIGURE 2.2. The San Andreas Fault clearly marks the boundary between the North American Plate (on the right) and the Pacific Plate (on the left) in Central California. © 2023 Google Earth, Maxar Technologies.

opposite sides of the fault creep in different directions. The city keeps patching up the streets and sidewalks, and homeowners annually repair their chimneys, try to level their doors, and repaint their houses. But the hidden forces just beneath the ground continue stretching Hollister like a rubber band. And the residents go right on patching and painting.

WHY EARTHQUAKES?

Over the past 55 years or so, geologists and geophysicists have developed a much clearer understanding of why earthquakes occur where they do. A glance at a map of worldwide seismicity quickly reveals that patterns are not random but that most earthquakes are concentrated along some very narrow zones. One of them follows the San Andreas Fault, almost the entire length of California.

The 1960s and 1970s witnessed a revolution in scientific thinking about the Earth and its history and evolution. Much of the initial evidence for the developing theories and concepts came from the exploration of the ocean basins, which began in earnest in the 1950s. The discovery of a world-encircling undersea ocean-ridge system that was volcanically active, a system of deep ocean trenches surrounding the

Pacific Ocean with associated chains of active volcanoes (the Ring of Fire), and a worldwide band of earthquakes that followed these unique features led to the development of the theory of global or plate tectonics in 1968. This all happened while I was finishing graduate school and starting in as a young assistant professor at the newly opened University of California, Santa Cruz campus and required a quick retooling of my own geologic understanding of the evolution of the Earth's surface and its major features.

There is now good agreement among Earth scientists that the Earth's crust (or lithosphere) is broken up into a number of large, rigid plates about 60 miles thick that are moving around relative to one another due to the motion of hot, partially fluid material within the underlying upper mantle (asthenosphere) of the Earth. Earthquakes are concentrated at the edges of these massive plates where they interact. Some of these plates are diverging, or moving apart, along a 25,000-mile-long oceanic volcanic mountain chain, which splits the middle of the Atlantic and Indian Oceans and crosses the eastern Pacific, looking a lot like the seam on a baseball—a very large baseball to be sure.

Earthquakes occur along these seafloor fissures where hot molten material forces its way up from the mantle to erupt and form undersea volcanoes. Thousands of miles away, the opposite sides of these plates form a second type of plate boundary as they collide. Something has to give at these head-on collisions, and deep trenches form where thin, dense oceanic plates are forced down beneath lighter and thicker continental plates. As the plates scrape against each other, large earthquakes frequently occur. Collision plate boundaries of this type occur almost completely around the margins of the Pacific Ocean, which is marked by an almost continuous series of deep, narrow trenches. The Earth's largest earthquakes take place at these subduction zones—offshore and underneath Alaska, Japan, the Philippines, New Zealand, Chile, Peru, Costa Rica, and Mexico, where oceanic plates descend beneath continental plates.

The San Andreas Fault is a third type of boundary, where two huge plates grind continuously but very slowly alongside one another (see figure 1.4). Similar faults occur in New Zealand and Turkey. The San Andreas Fault zone, with its associated branches and splinters, is up to 50 miles wide in the greater San Francisco Bay area. This zone separates the North American Plate, which extends east all the way to the middle of the Atlantic Ocean, from the Pacific Plate, which consists of almost half of the Earth's surface, the entire Pacific Ocean extending clear to Japan. This is a huge piece of real estate, and these two plates just happen to rub against each other throughout 700 miles of California landscape extending from the Mexican border to Cape Mendocino in Northern California (see figure 1.4). They are moving alongside each other at about two inches a year. There isn't a smooth lubricated surface between them, however, so they tend to stick together until decades of strain accumulates and the rocks finally rupture, with the plates moving more or less instantly in opposite directions. The break in the rocks initiates an earthquake, which sends seismic waves out in all directions at several miles per second. This unfortunately makes calling someone to warn them of an impending quake after you felt it of somewhat limited value.

If we accumulated two inches of strain each year along the San Andreas Fault, in 50 years we would have the potential for an earthquake that could produce up to 8 feet of horizontal displacement; in 100 years, the accumulating strain could generate over 16 feet of rupture, and so on. Based on the rate of movement along the fault, the approximately 200-year historic earthquake record, and excavations into the soils and sediments along the fault where we can recognize evidence for large prehistoric earthquakes, we now know that large earthquakes (magnitude 7 or greater) occur about every 50 to 100 years along the San Andreas Fault system. The entire fault doesn't all rupture at once, however, and individual segments of the fault can behave differently.

Over the approximately 15–20 million years of the San Andreas Fault's existence, the part of California west of the fault, extending from San Diego to Cape Mendocino and including most of California's coastal counties and their residents, has moved northwest about an inch and a half per year relative to the rest of the state to the east (see figure 1.4). Thus, California is very slowly being stretched and torn apart, and the small earthquakes that occur daily, as well as the larger shocks that occur every 10 to 20 years, are evidence of the constant grinding, or friction, along the boundary between these two massive slabs of the Earth's crust. Assuming the present plate motion continues, western California will eventually drift off, albeit very slowly, and become the Madagascar of the North Pacific, drifting towards Alaska.

The Parkfield segment of the fault, which is about 15 miles northeast of Paso Robles in Central California, had slipped every 20–25 years throughout the past century, producing earthquakes in the 6.5 magnitude range. Earthquakes had occurred in 1857, 1881, 1901, 1922, 1934, and 1966, not quite like clockwork but fairly regularly.[3] This Parkfield seismic history made it seem like a great opportunity to instrument and monitor the area to see if we might learn something about earthquake precursors. Because the last earthquake had shaken the Parkfield area in 1966, United States Geological Survey (USGS) scientists felt the next one was due about 1986 and decided to place a dense array of instruments along this segment of the fault. Their objective was to record any possible precursors to the next quake, to get a clearer idea of whether any measurements or observations might prove useful for predicting future earthquakes elsewhere, something that has continued to elude seismologists. Despite the historical regularity and all of the instruments that had been installed, the pattern didn't continue, and it wasn't until September 28, 2004, 38 years after the previous quake, that a magnitude 6.0 earthquake finally rocked Parkfield. Earthquake prediction has proven to be almost completely elusive so far, not only in California, but globally as well.

WHAT HAPPENS DURING AN EARTHQUAKE?

Although many early civilizations, as well as our own Native Americans, had their beliefs and legends about the causes of earthquakes, it wasn't until the great San Francisco quake of 1906 that we first understood the significance of the San Andreas Fault. That event showed us that it is breakage or displacement along a fault that produces the intense ground shaking we feel and that can destroy buildings and bridges and claim lives. As the rocks rupture or break, the strain that has accumulated for decades is released as seismic waves. These waves radiate out away from the zone of rupture, much like the ripples that radiate outward from a pebble thrown into a pond. There are several different kinds of waves that are generated, but they all shake the ground and they all can cause damage; the stronger and longer the shaking, the greater the damage is likely to be.

In general, the magnitude or size of an earthquake, the distance from the rupture zone, the type of rock or soil we build on, the materials we build with, and the quality of construction are the factors that affect the damage to be expected during a quake in any particular location. Everything else being equal, which is rarely the case, the larger the earthquake and the closer we are to the rupture zone, the greater the shaking and damage we can expect.

As we move from hard crystalline rock like granite, to softer sedimentary rocks like sandstone or shale, to unconsolidated or loose sediment like silts and clays, and finally to water-saturated sediments like river bottom, estuary, or shoreline sediments, we can expect the damage during an earthquake to substantially increase. This has been borne out during all of the historic earthquakes in California and was first noticed in 1906 following the great San Francisco earthquake.

Much of the historic development around the margins of San Francisco Bay occurred on unstable ground, often artificial fill, which became painfully clear in both the 1906 San Francisco and 1989 Loma Prieta earthquakes.[4] Reports on the great San Francisco earthquake documented the exaggerated shaking and consistently greater damage to buildings in the lower waterfront areas of the city, which were underlain by the thickest bay mud and filled land, as compared to those buildings on the higher bedrock hills. Moreover, the nearby cities of Salinas, San Jose, Palo Alto, and Santa Rosa, built on deep alluvial soils, suffered to an extent far out of proportion to their distance from the epicenter in 1906. Again, in 1989, structural damage in San Francisco's Marina District and the tragic collapse of the Cypress Freeway overpass in Oakland, which was responsible for two-thirds of the deaths in the earthquake, took place 65 miles from the epicenter and was due to the amplified seismic shaking in the weak underlying sediments (see figure 2.1). In Southern California, the San Fernando Valley suffered heavy losses in both the San Fernando earthquake in 1971 and the Northridge earthquake in 1994 for the same reasons.

Seismic shaking is the single greatest hazard during a large earthquake and can affect thousands of square miles and have major impacts many miles away. Ground breaking or surface rupture along a fault will also typically occur during large earthquakes. In the 1906 San Francisco event, ground breaking was widespread and allowed for the identification and mapping of the trace of the San Andreas Fault. These offsets or ruptures don't normally present a threat to human life but can be damaging or disastrous to structures built directly across faults. While some very large cracks opened up in the Santa Cruz Mountains during the 1989 Loma Prieta earthquake, these didn't swallow up people or animals as some early legends and sketches portrayed.

Ground cracking is a common earthquake effect, whether due to movement along the fault itself or to landsliding, settling, or liquefaction. Maximum horizontal offset along the San Andreas Fault in 1906 reached as much as 20 feet near Point Reyes, north of San Francisco. In Santa Cruz County the greatest offset documented was about 6 feet in a railroad tunnel beneath the Santa Cruz Mountains.[5] Displacement of this sort along a fault is usually confined to a relatively narrow zone, but when the fault ruptures, there is not much we can do to control it, other than to completely avoid the fault zone to begin with. Because fault zones may be hundreds of feet wide and don't always break in the same place, it is nearly impossible, even for geologists, to know precisely where or when the fault may rupture next.

Slope failures during severe seismic shaking can produce significant damage. The type of failure depends upon the nature of the rock or soil and the water content. The winter of 1905–06 had been very wet, such that the April 18, 1906, quake hit when many soils were saturated, and as a result, the earthquake initiated large landslides, earthflows, and mudflows over the greater San Francisco Bay area.[6] In contrast, the October 17, 1989, Loma Prieta earthquake struck after an extended dry period. Older dormant landslides in the Santa Cruz Mountains were disturbed enough to move a few feet and open up cracks that passed under and damaged roads and houses, but no earthflows occurred.

Where sandy soils are saturated, or all of the pore spaces between the grains are filled with water, there can be a high potential for liquefaction with prolonged seismic shaking. Think of quicksand, or the process you can observe on a sandy beach when you stand right at the water's edge and jiggle your feet around in the wet sand. The disruption of the grains increases the pressure of the water between the grains and actually leads to liquefaction, or turning the wet sand into a fluid. During large earthquakes this process may be widespread along sandy shorelines, on river bottoms, or in alluvial soils with high water tables. As sandy soils turn to liquid, they lose the ability to support heavy overlying structures, and the water and sand can actually erupt at the ground surface and create small sand volcanoes. Failure of these same unstable soils during shaking can also produce what has been termed lurch cracking along riverbanks and other slopes. This was common in both 1906 and 1989 along the Salinas, Pajaro, and San Lorenzo Rivers and their floodplains.[7]

What is important to understand if we are to reduce damage and destruction from future earthquakes is that we need to learn from the disasters of the past. The same kinds of processes tend to affect the same areas repeatedly. The detailed reports from the early geologists who traversed California on foot and horseback after the 1906 earthquake were strikingly familiar to the newspaper accounts and geologic investigations of what we witnessed in 1989. While we cannot predict precisely when future earthquakes will occur, we do have a pretty clear idea of where they will occur and how specific areas will be affected. This isn't a mystery and it isn't rocket science; it's just learning from the observations of the past.

EARTHQUAKE HISTORY OF CALIFORNIA

While hundreds of small to moderate earthquakes are felt each year in California, what follows is a description and some accounts of the larger historic shocks as reported by various sources. Not surprisingly, the further we go back in time, the greater are the uncertainties and the less accurate is the information that's available. Seismographs did not come into common use until the early 1900s and were followed by the development of the first magnitude scale by Charles Richter in 1935. Prior to this time, earthquakes were described by the intensity of shaking, as observed by people who experienced the earthquake firsthand, by using the Modified Mercalli intensity scale (table 2.1). The intensity of any individual earthquake will vary widely from place to place, however, depending upon the distance to the epicenter and the surface materials. An intensity of VI or higher is indicative of a moderate to large earthquake with a high likelihood of some damage, and when we reach VIII, IX, or X, damage can be widespread and severe.

January 9, 1857

The 1857 7.9 magnitude Fort Tejon earthquake was the last really big one in Southern California, comparable in size to the great 1906 San Francisco earthquake to the north. The San Andreas Fault ruptured in 1857 over a distance of about 225 miles, somewhat shorter than the nearly 300-mile-long break in 1906. Maximum displacement along the fault, however, was nearly 30 feet, compared to approximately 20 feet in 1906. The major difference in these two large shocks was the level of development and population in the two areas. While there was an army post at Fort Tejon, which was damaged, there was little else in close proximity at that early date in California's history, compared to San Francisco in the early years of the 1900s.

Shaking lasted for one to three minutes and was felt from Marysville, north of Sacramento, to San Diego and as far east as Las Vegas, Nevada. Changes in the flow of streams and springs were recorded in San Diego, in Santa Barbara, and at the south end of the San Joaquin Valley. Ground fissuring and hydrologic changes were reported from Sacramento to the Colorado River delta and in the beds of the Los Angeles, Santa Ana, and Santa Clara Rivers. The 1857 earthquake was a major event, although it only resulted in a single fatality due simply to the fact that it struck in a very sparsely populated area. There is now 166 years

TABLE 2.1   MODIFIED MERCALLI EARTHQUAKE INTENSITY SCALE

Intensity	Description of Shaking/Damage
I	Not felt except by a very few under especially favorable conditions.
II	Felt only by a few persons at rest, especially on upper floors of buildings.
III	Felt quite noticeably by persons indoors, especially on upper floors of buildings. Many people do not recognize it as an earthquake. Standing motor cars may rock slightly. Vibrations similar to the passing of a truck. Duration estimated.
IV	Felt indoors by many, outdoors by few during the day. At night, some awakened. Dishes, windows, doors disturbed; walls make cracking sound. Sensation like heavy truck striking building. Standing motor cars rocked noticeably.
V	Felt by nearly everyone; many awakened. Some dishes, windows broken. Unstable objects overturned. Pendulum clocks may stop.
VI	Felt by all, many frightened. Some heavy furniture moved; a few instances of fallen plaster. Damage slight.
VII	Damage negligible in buildings of good design and construction; slight to moderate in well-built ordinary structures; considerable damage in poorly built structures; some chimneys broken.
VIII	Damage slight in specially designed structures; well-designed frame structures thrown out of plumb. Damage great in substantial buildings, with partial collapse. Buildings shifted off foundations.
IX	Damage considerable in specially designed structures; well-designed frame structures thrown out of plumb. Damage great in substantial buildings, with partial collapse. Buildings shifted off foundations.
X	Some well-built wooden structures destroyed; most masonry and frame structures destroyed with foundations. Rails bent.
XI	Few, if any (masonry) structures remain standing. Bridges destroyed. Rails bent greatly.
XII	Damage total. Lines of sight and level are distorted. Objects thrown into the air.

NOTE: Re-created from Griggs and Gilchrist, 1983, 34.

of strain that has accumulated along this stretch of the fault, however, and there are considerably more people in the area, so another major earthquake will undoubtedly have far greater impacts.

October 8, 1865

"The most severe shock since the annexation of this territory" occurred on this date in 1865.[8] Recent reanalysis indicates that this earthquake was probably centered in the Santa Clara Valley and the eastern foothills of the Santa Cruz Mountains, and perhaps not on the San Andreas

Fault. It is thought to have been one of the five largest quakes to strike the San Francisco Bay region in historic times. The magnitude of this quake has recently been estimated at 6.5, so a little smaller than the 1989 Loma Prieta event. This earthquake caused damage from San Juan Bautista in the south to Napa in the north and ruined the city hall in San Francisco as well as damaging numerous water and gas pipes.

The initial reports from Santa Cruz stated that "every brick building here is ruined."[9] Total losses in the city were estimated at $10,000 (this statement suggests that the number and worth of the city's brick buildings were not terribly high in 1865). The ground settled along the San Lorenzo River, cracking the soils along its banks. Some small cracks emitted jets of water two to four feet high for several minutes, indicating liquefaction in the subsurface. Near Soquel, the sea was rising and falling with what some described as "convulsive throbs," carrying some of the high cliffs into the sea (presumably in the vicinity of Capitola at that time).

In Watsonville, there was a grand total of $2,000 in damage, $1,500 of which was at the Pajaro Flouring Mills. Well-constructed buildings or those built on solid ground suffered little or no damage. Cracks appeared on the banks of the Pajaro River, ranging from 10 to 15 inches wide and hundreds of yards long. Later, observers remarked, "It is a singular fact that the shock was most severe at Santa Cruz and along the lower part of the Pajaro River."[10] This statement is a clear indication that residents noticed 150 years ago that the river bottomland with deep sediments and a high water table (like downtown Castroville, Watsonville, and Santa Cruz) would experience more intense seismic shaking than areas on solid ground or bedrock. These same patterns have been repeated in virtually every subsequent large earthquake in Central California.

Highest intensities during this event were felt in the mountains between Santa Cruz and San Jose. At Mountain Charlie's on the old Santa Cruz Road, the earth opened up in several places, and steam and water were emitted from the cracks. On the Santa Cruz Gap Road, chimneys were thrown down, and the roads were more or less obstructed by boulders that rolled from the hillsides. Wells and streams in Santa Cruz County were markedly affected as many of their volumes doubled. Water also boiled up from the ground for half an hour after the shock. This was observed at the old Santa Cruz mission orchard, which was presumably down on the floodplain.[11]

October 21, 1868

One hundred and fifty-five years ago, in the early autumn morning of October 21, 1868, an earthquake with an estimated magnitude of 6.8 shook the greater San Francisco Bay area for about 40 seconds. This shock was known widely until the great 1906 event as "the San Francisco earthquake." Surface rupture was recorded along about 20 miles of the Hayward Fault on the east side of San Francisco Bay where damage was extensive. Six feet of horizontal displacement was reported to have taken place, and although population was low at the time, losses were great, with almost every building in the small town of Hayward damaged to one degree or another. In San Leandro, which only had a population of 400 at the time, the second floor of the Alameda County Courthouse collapsed (figure 2.3). Damage also occurred in Oakland, San Francisco, San Jose, Napa, and Santa Rosa, with a total loss of 30 lives.[12] It still ranks as one of the state's more destructive earthquakes, and the fault will rupture again, maybe soon.

FIGURE 2.3. Partial collapse of the Alameda County Courthouse during the 1868 earthquake. *Courtesy of Hayward Area Historical Society, public domain.*

Trenching into the subsurface along the Hayward Fault in recent years has led to the recognition of evidence for 12 large historic earthquakes over the past 1,900 years. The time between each of these earthquakes has varied between 95 and 183 years, with an average of about 150 years.[13] With the 1868 quake having taken place 155 years ago, USGS scientists believe that the fault has reached the point where a powerful damaging earthquake could occur at any time. Following the 1868 shock, and an earlier one in 1865, engineers developed methods to strengthen buildings in the San Francisco Bay area. Brick and other masonry buildings were retrofitted with iron tie-rods and anchors between floors and walls so that buildings could better withstand strong ground shaking. The greatest improvement came in 1885 with the beginning of steel-frame buildings, which proved very effective in those structures built in San Francisco before 1906.

The Hayward Fault splits off from the Calaveras Fault northeast of San Jose and extends northwest along the east side of San Francisco Bay for nearly 60 miles. It passes beneath Fremont, Hayward, San Leandro, Oakland, and Berkeley before plunging off into Suisun Bay. The fault passes directly under UC Berkeley's Memorial Stadium, among many other public buildings (figure 2.4).

The Hayward Fault has often been labeled as the nation's most dangerous fault for two reasons. First, it is the most urbanized fault zone in the United States. While there were just 24,000 residents in Alameda County in 1868, by 2023 there were 1.73 million. Accompanying the population growth are hundreds of homes and other structures built along the trace of the fault, as well as freeways, mass transit corridors, gas and water pipelines, and electrical transmission lines that cross the fault. Second, the fault has a history of repeated earthquakes on average about every 150 years, and the last big one occurred 155 years ago.

April 18, 1906—The Great San Francisco Earthquake (and Fire)

The magnitude 7.9 1906 earthquake, which unzipped nearly 300 miles of the San Andreas Fault, from Shelter Cove, near Cape Mendocino, to San Juan Bautista, east of Monterey Bay, was the most destructive to Central California in historic time and also ranks as one of the most significant earthquakes ever recorded. Its significance comes from the massive amount of information collected by scientists after the shock. The event, with its great rupture length and large horizontal displacement, opened the eyes of those geologists who studied its effects to the

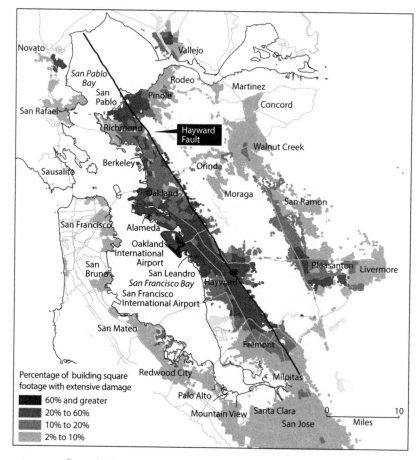

FIGURE 2.4. Potential damage from a 7.0 magnitude earthquake on the Hayward Fault. *Courtesy of US Geological Survey.*

importance of this crack across the landscape, which wouldn't be completely understood for another 62 years, when the concept of plate tectonics first emerged as a unifying theory.

Although the epicenter is now believed to have been just off the Golden Gate, the shaking was felt from Coos Bay, Oregon, to Los Angeles and east to central Nevada, an area of about 350,000 square miles. The 1908 Carnegie Commission report on the earthquake provided a very thorough description of its effects throughout California by a group of dedicated geologists who traversed the region, primarily on horseback, to document the extent of the ground failure and damage. Some selected observations are included below, in some cases exactly as written (indented as extracts). This was the event that made all of California, including its geologists, aware of the role and significance of the San Andreas Fault.

The impacts of the shaking from this shock have turned out to be good predictors of where damage would occur in future large earthquakes, such as Loma Prieta in 1989. One major conclusion reached in the Carnegie Commission report was that the intensity of shaking was closely correlated with the underlying geologic materials. Those areas and communities situated in sediment-filled valleys with deep floodplain or alluvial soils (Santa Rosa, Palo Alto, San Jose, Santa Cruz, Watsonville, and Salinas for example) sustained stronger and more damaging shaking than the higher areas underlain by bedrock (the "low ground" vs. the "high ground"). The shaking lasted as long as 42 seconds, and the maximum shaking intensity reached XI (extreme—see table 2.1) in San Francisco and areas to the north, such as Santa Rosa.

Within the city and county of San Francisco, the investigation clearly demonstrated that the amount of damage produced by the earthquake of April 18 in different sections of the city and county depended chiefly upon the geological character of the ground. In areas underlain by solid rock, the earthquake produced little damage, whereas on what was referred to as made land, such as artificial fill on the margins of the bay, shaking was more intense, lasted longer, and produced much greater damage.

More than 3,000 people are believed to have died from the earthquake, and over 80 percent of the city of San Francisco was destroyed from the combination of the shock and the subsequent fires (see figure 1.2). This death toll remains, over a century later, the second-largest number of fatalities from a natural disaster in California's history and also ranks high on the lists of the nation's natural disasters. The great floods of 1861–62 rank as number one, with a best estimate of 5,000 fatalities (see chapter 5). Out of a San Francisco population of about 410,000, somewhere between 227,000 and 300,000 people were made homeless, with many of these fleeing across the bay to Oakland and Berkeley and many camping out in Golden Gate Park and the Presidio for months.

While the earthquake and its aftershocks were devastating, the fires that were caused by ruptured gas mains and that burned out of control for four days and nights afterwards were more damaging. Estimates were that up to 90 percent of the losses were from the fires, and approximately 490 city blocks and 25,000 buildings were destroyed. Some of the fires were caused by the untrained use of dynamite by the fire department in an effort to demolish buildings and create firebreaks. Property losses reached the equivalent of about $10.1 billion (in 2022 dollars).

While entire volumes have been written about the earthquake and its effects, with the most comprehensive being *The California Earthquake of April 18, 1906: Report of the State Earthquake Investigation Commission,* the pages that follow are intended to provide some overall perspective of the magnitude and impacts of the great 1906 earthquake in some of the state's most affected areas.[14] Where passages have been taken directly from this report, they are indented as extracts.

Humboldt, Mendocino, Sonoma, and Marin Counties

Despite the distance to the epicenter (~275 miles), strong shaking and damage were reported as far north as Eureka.[15] The quote below refers to the town of Ferndale:

> This town, on the south side of the flood plain of the Eel River, appears to have been the most severely shaken place in Humboldt County . . . not a chimney was standing and every brick building was torn to pieces. All of the plate-glass windows were smashed. . . . In Petrolia the shock threw every house off its foundation. . . . The houses are built on the soft bottom land of the Mattole River, several of them within a few feet of the river, and their supports are simply blocks of wood, stone, or concrete resting on the surface of the ground . . . in the mountains it opened great fissures, ruining many acres of good grazing land.

Fort Bragg

Along this stretch of California coast, the San Andreas lies just offshore.

> The town of Fort Bragg suffered quite severely, and the indications are that the intensity of the shock was considerably greater than in the towns immediately to the south. Several brick buildings were completely demolished; others had parts of their walls broken off. Even a number of wooden buildings collapsed or were partially wrecked. Fire broke out and devastated 1½ blocks before it could be controlled.

Mendocino

> This town, like Fort Bragg, is on the first of a series of wave-cut marine terraces. . . . The town shows little damage. Only one large frame building, the Occidental Hotel, was wrecked thru the giving way of its underpinning. Few chimneys escaped destruction.

Point Arena

All the brick buildings in the place have completely collapsed, and in the opinion of the residents it was deemed wisest to replace them with frame [wooden] structures. All brick chimneys had fallen. As a result of the shock, fire started in the chemical laboratory of the grammar school, and that building, together with the Methodist Church adjoining it, burned down. The Point Arena lighthouse, 3 miles west of the fault, was thrown out of the vertical . . . it has been condemned as unsafe and is to be torn down.

Population density was quite low along the trace of the fault through the counties north of San Francisco in 1906, so the effects of the earthquake were primarily surface rupture along the trace of the fault; offset or displacement of fences, roads, or other features; and the collapse of or damage to scattered farmhouses and barns. The greatest overall displacement documented during the quake was near Olema in the Point Reyes area and reached about 20 feet.

Healdsburg

This place comes next to Santa Rosa in the extent of damage done to towns in Sonoma County. The new 3-story brick building of the Odd Fellows Society is a total wreck, as are several other buildings, but many brick structures stood the shock without serious damage.

Tomales

The Catholic Church, a fine-looking stone building, was completely wrecked, as were the brick bank and saloon, and a stone store building. Several frame buildings were pitched from their foundations and wrecked. . . . All chimneys were down.

Santa Rosa

This city lies on the eastern side of Santa Rosa Valley, which is an alluvial plain underlain by hundreds of feet of unconsolidated sediments.

This city, with a population of 6,700, suffered relatively more than any other place in California, except perhaps Sebastopol and Fort Bragg.

The earthquake and following fire (from gas line breakage) led to at least 61 fatalities and essentially destroyed the business portion of Santa Rosa, including the county courthouse (figure 2.5).

The residence portion of the town suffered to quite an extent. Chimneys were generally thrown down or so badly cracked so as to necessitate their

FIGURE 2.5. Collapse of the Sonoma County Courthouse in Santa Rosa in the 1906 earthquake. *Courtesy of Steinbrugge Collection, NISEE-PEER, University of California, Berkeley, public domain.*

rebuilding. From twenty to twenty-five residences were thrown to the ground by the collapse of their underpinning, and badly wrecked.

Sebastopol

Several buildings were completely wrecked. The 2-story Knowles Hotel, a frame building, veneered with brick, went completely down, flattening the first story. The walls of the hotel fell out, so that the occupants of the rooms in the second story walked out on the ground level. . . . Three stores just south of the post-office were completely wrecked.

San Francisco and San Mateo Counties

The rupture zone followed the trace of the San Andreas Fault through San Andreas Lake and Crystal Springs Reservoir, adjacent to today's Highway 280. Large water pipes were broken, and fences and roads were horizontally offset by more than 10 feet along the path of the fault in this valley.

Stanford University

The center of the university lies about 4.3 miles from the San Andreas Fault. Most of the university is constructed on the alluvial sediments of the Santa Clara Valley and, as a result, suffered sustained shaking and considerable damage.

> There were 61 residences on the campus of Stanford University at the time of the earthquake. Out of 140 chimneys on these buildings, 104 were thrown down. . . . A frame building occupied by the Chi Psi Fraternity was so badly wrecked that it had to be abandoned. The injury done to this building was due to its having stood upon posts 4 feet high and not well braced; the swaying of the building threw it off these supports. . . . The Stanford residence . . . was so badly wrecked that it has since been torn down.
>
> Of the University buildings proper, some were unhurt while others were completely wrecked. They all stand upon the gravelly loam of the Santa Clara Valley floor. As a rule, the older the buildings were, the better they withstood the shock. At Encina Hall, the men's dormitory, one chimney fell thru the roof and carried down a tier of rooms into the basement, killing one student. The south ends of the wings of Encina Hall were so badly cracked that they had to be completely rebuilt. The chimneys also fell from Roble Hall, the women's dormitory, and did some damage to the roof and upper floors; the building, which is of concrete, was otherwise unhurt. The Chemistry building had 32 tile-lined stone ventilating chimneys projecting 12 to 16 feet above the roof . . . these were all thrown down. The Geology building, at the southwest corner of the outer quadrangle, was the last building of this group to be put up. It was a 3-story structure and has barely been finished; but it was not yet occupied when the earthquake occurred. Sections of the walls were thrown down from every facet of the building. These sections extended from the eaves down to the second floor. The new gymnasium, a stone-faced building, was totally wrecked. It has just been put up. . . . The new library, also a stone-faced brick structure was completely wrecked except a tower of steel on which its central dome still stands. The Museum building consisted of an older central portion built of concrete, and extensive additions of brick had just been completed. The new brick portions were almost all thrown down, but the older concrete was unhurt.

Santa Clara County

San Jose

> The earthquake threw down many brick and stone buildings, and with the exception of 4 or 5, damaged all the rest of the brick buildings, more or less. The damage done to frame houses was proportionally less. Forty buildings were counted, however, that were thrown off their foundations. . . . In many instances these buildings were completely demolished. Numerous wind-mills and [water] tanks capsized, while at least 95 percent of the brick chimneys throughout town fell.

Milpitas

Nearly all chimneys were here thrown down. . . . There are no brick buildings in the village and the destruction seems insignificant.

Agnews

The insane asylum, consisting of three tall and three minor brick buildings and some small frame structures, suffered very severely. Every one of the brick structures was damaged beyond repair and will have to be completely rebuilt. The main buildings were long, 3-story brick structures oriented north and south, with large projecting bay windows at their north and south ends. These were destroyed, so that both buildings are open at their ends. The fall of these walls caused the caving in of the roof, and the sagging down in some places of floors. Numerous lives were lost; in all 112 dead being found in the ruins. The administration building was partly wrecked by the fall of its tower, which crashed thru the roof and all the floors, carrying with it a number of people. The extent of the destruction is in some measure due to the use of weak mortar, the bricks having, as a rule, fallen separately rather than in aggregates. It is believed that well-built buildings would not have suffered such wholesale destruction as was witnessed here.

Santa Cruz County

Santa Cruz

The city of Santa Cruz furnishes excellent evidence on the effect of soil formation on the intensity of the earthquake shock. On the high ground in Garfield Park and also in the northwest part of the city, only about one-fourth of the chimneys fell and a little plastering was cracked, while in the lower ground near the business section, several brick and stone buildings were partly shaken down. The San Lorenzo River was churned into foam, the banks cracking and settling several inches, and sand, said to have come from a depth of 100 feet, was forced up in several places.

The courthouse (the old Cooper House) on Pacific Avenue was almost destroyed as the cupola fell through the ceiling and landed in the basement. Plate glass windows broke along Pacific Avenue, and at least one brick building collapsed. It was estimated that one-third of the chimneys in the city were either destroyed or damaged, and all of the bridges across the river were reported to be badly wrenched and declared unsafe. Cracks opened up in the street near the railroad depot and at the corner of Front Street and Soquel Avenue. Broken water mains and the eight-inch city water pipe at Wilder's dairy were broken and twisted, which shut off the city's water supply. All telegraph lines between Santa

Cruz and points north were thrown down, keeping the city uninformed of the destruction of San Francisco for two days.

Capitola

> Nearly all of the chimneys at Capitola fell, and considerable plaster was shaken from the north walls of the first floor of the hotel. . . . Much earth fell from the bluffs [probably Depot Hill] near the town, but there was no appreciable effect on the surf. At the country bridge across Soquel Creek, the ground at the east abutment moved inward, cracking the concrete and buckling a water pipe . . . a continuous cloud of dust rose along the cliffs between Castro's Landing [now called Rio Del Mar] and Santa Cruz.

Soquel

> In the low ground at Soquel, nearly all of the chimneys fell, but most of those on high ground stood. Much plaster fell and goods were thrown from shelves in the business section, which is close to the creek.

Watsonville

Watsonville is built over the loose alluvium of the Pajaro River Valley, where seismic shaking is amplified. The city suffered even more damage than Santa Cruz.

> About 90 per cent of the chimneys were broken off at the roof-line, the greater portion being near to the river. . . . Parts of a few brick walls near the river fell, and considerable settling of the ground took place in Chinatown on the southern side of the river.

Summit Ridge

> At Summit, a summer resort, the new hotel and several small cottages were all thrown toward the north. The main fault fracture is about 500 feet northeast of the hotel, and a secondary crack close to it has a downthrow of from 5 to 7 feet on the north or downhill side. . . . The Summit schoolhouse was dropped 4 feet downhill from its original position. . . . All brick chimneys on the ridge fell, mostly to the north. . . . The banks of Burrell Creek appear to have approached each other, so that the creek has become very much narrower.

This was no doubt due to landslides or hillside failures, which moved material downslope into the stream bottoms.

> The Morrell ranch is located 1 mile south of Wright's Station and is on the line of the fault. The house itself was exactly upon a fissure, which opened up under the house at the time of the earthquake. The house was completely wrecked, being torn in two pieces and thrown from its foundation. . . . At

Freely's place, 4 or 5 miles north of Morrell's, some 15 acres of woodland have slid into Los Gatos Creek, making a large pond.

Perhaps the most tragic and devastating events in the county were the landslides and mud and debris flows resulting from the earthquake. Although there are many similarities in the effects of the 1906 and 1989 earthquakes on the lowlands, the effects in the mountains were vastly different due to the differing rainfall patterns prior to the quakes. The April 1906 quake followed a period of above-average rainfall, whereas the 1989 shock came after three years of drought. The excess water in the soils at the time of the 1906 quake led to large landslides and debris flows, which moved considerable distances downslope.

On Hinckley Creek, a tributary of Soquel Creek, a landslide 500 feet wide extended all the way to the ridgetop and descended with "extraordinary speed," burying the Loma Prieta lumber mill under a mass of rock and trees.

> The mill, boarding house, and other buildings of the plant were situated in a gulch and were overwhelmed by a portion of the mountain—1500 feet long, 400 feet wide and 100 feet deep, which slid down on top of them. The mill and everything in the gulch were forced up the opposite slope of the mountain and there buried to a depth of one hundred feet.

Nine men were buried instantly, while others, only several hundred feet away, were spared. The landslide dammed the stream, forming a lake up to 100 feet deep. Hundreds were involved in a massive digging effort in the following week, but only three bodies had been discovered after five days of searching. More than a year passed before the last body was recovered.[16]

Monterey County

Monterey, Pacific Grove, and Carmel are all built almost entirely on the granite of the Monterey Peninsula, and as a result, damage on the peninsula was very minor during the 1906 earthquake.[17] They are also farther from the San Andreas Fault and the 1906 rupture zone. Only several chimneys were damaged in Pacific Grove, although shaking was moderate to severe according to residents. The Pacific Grove lighthouse was built on sand dunes and suffered greater damage as a result, with cracking of the dome structure. Monterey experienced essentially the same intensity of shaking as Pacific Grove, with no apparent damage done to houses, and the only losses reported were some broken glassware in stores and some top-heavy furniture overturned.

The Del Monte area suffered to a greater degree, particularly the Del Monte Hotel, which was built on alluvium and fill and was surrounded by marshy land, ponds, and sand dunes.

There were over 50 chimneys in the hotel, and half of them were thrown down, one crashing thru the roof on the west side of the hotel and causing two fatalities. The chimneys were tall and top-heavy, having ornamental tops; and while the damage to the interior of the hotel was very slight, showing that the earthquake was not of a violent type, the vibrations were sufficient to throw these top-heavy chimneys.

In the fields between Monterey and Castroville, geysers were reported that extruded boiling hot, bluish . . . mud to a height of ten to twelve feet. The railroad tracks for almost the entire distance are twisted and lost all semblance of tracks. . . . Near Castroville, while the disturbance was at its height, Foreman H. J. Hall grabbed his two children . . . and as they passed through the door, they saw the earth open up and a crevasse, which Hall described as fully six feet wide, open and close several times.

An observer that evening noted:

the house standing in a pool of geyser mud . . . like quicksand, and of unknown depth.

All of these descriptions speak to the widespread process of liquefaction during the 1906 event in the lowlands along the Salinas and Pajaro Rivers.

Buildings and facilities out on the Moss Landing sand spit were extensively damaged due to the unconsolidated nature of the underlying sand and a high water table.

The wharf at Moss Landing buckled up and partly collapsed, while the warehouses were wracked and fell westward . . . the condition of the wharf indicates an eastward movement of the sand-spit. It is reported that at places along the pier where the water was formerly 6 feet deep, it now has a depth of 18 or 20 feet.

This no doubt was a result of slumping of sediments into the head of Monterey Submarine Canyon, which also happened in 1989.

Where there were sand dunes a few days ago, now there are deep holes with water bubbling through them. . . . At the hotel and stores on the mainland [at Moss Landing], brick chimneys fell, but plastering was not seriously cracked. . . . At Moss Landing, where the [Salinas] river runs parallel with the shoreline, the strip of land is seamed for miles. A crack, or rather a sink, about 20 feet wide and 4 or 5 feet deep ran under the buildings and rent them asunder. The office building between this crack and the river has been moved bodily—land and all—about 12 feet toward the river. Some of the cracks run into the ocean.

Until 1910 the lower course of the Salinas River flowed north, behind the sand spit through present-day Moss Landing Harbor, past the entrance to Elkhorn Slough, and discharged into the ocean north of the present Moss Landing Harbor entrance.

The marshlands, riverbanks and some farmland along McClusky Slough and along the Pajaro River near its juncture with the Salinas River north of Moss Landing were extensively cracked. . . . Fresh water came out of some fissures. The bottom of the Pajaro River came up to a point just north of its juncture with the Salinas River causing Pajaro River to change its course and empty into Monterey Bay near the present mouth of the river.

Despite being 100 miles from the earthquake epicenter, the towns along the Salinas River valley suffered from severe seismic shaking, as did those valley floor communities to the north, such as San Jose and Palo Alto, which were also built on deep alluvium.

The town of Salinas suffered greater damage than any other place in the county. Nearly every house and building was damaged to some extent. Plaster fell, windows broke, chimneys fell or were cracked, and brick buildings had their upper portions thrown off and, in some cases, almost completely demolished. The town is on the flat valley land, about 3 miles east of the river . . . the flood plain of the Salinas River was caused to lurch toward the stream from both sides . . . these have the effect of landslide scarps and terraces . . . numerous craterlets were formed by the sudden ejection of water from the underlying sands [liquefaction] due to the compressive action of the shock. [Near Spreckels] water gushed forth at numerous places . . . and spurted repeatedly as high as 20 feet . . . for 10 minutes after the shock.

The Spreckels sugar mill was a very large (500 feet long, 150 feet wide, and five stories high) steel and brick building on the south side of the river. The whole structure was shortened along its long axis, walls buckled and bulged, the ground outside heaved and deformed, and water gushed from the ground. Damage was extensive. Deformation, settlement, and lurching of river bottom sediments continued all the way down the Salinas Valley to King City, where the riverbed sank nearly 6 feet.

June 29, 1925—Santa Barbara

The 1925 Santa Barbara earthquake hit about a century ago at 6:42 a.m. and produced shaking that lasted for about 18 seconds. Its magnitude was determined to be between 6.5 and 6.8, and shaking was felt from Paso Robles in the north to Santa Ana in the south. The source of the earthquake was located offshore in the Santa Barbara Channel along what appears to have been extension of either the Mesa–Rincon Creek fault or the Santa Ynez Fault system.

The twin towers of Mission Santa Barbara, which had stood since 1786, partially collapsed, and about 85 percent of the downtown buildings were damaged or destroyed. Unreinforced masonry, mainly brick, concrete, and stone construction, was discovered to suffer greatly during seismic shaking. Most homes survived the earthquake in relatively good condition, although most chimneys fell. A dam in the hills above the city collapsed, sending 45 million gallons of water through the streets. Somewhat surprising, considering the collapse of so many buildings, was that there were just 13 fatalities. In large part this appears to have been due to several heroes who shut off the city's gas and electricity, which no doubt kept fires from igniting. Businesses opened out in the streets, and many of the residents spent much of the summer sleeping outdoors as aftershocks continued.

The central part of the city was originally built in the Moorish Revival style, but because of the extensive damage (figure 2.6), the decision was made to rebuild in the Spanish Colonial Revival style, which has continued to the present. The city building codes were strengthened following the earthquake, due to the failure of so many downtown buildings that had been constructed of unreinforced concrete, brick, and masonry. Unfortunately this lesson had to be repeatedly learned in other California cities with subsequent earthquakes.

October 22, 1926—Monterey Bay

Two earthquakes of estimated Richter magnitude 6.1, centered in Monterey Bay, shook the central coast in October 1926 and were felt from Cloverdale in Sonoma County to Lompoc in Santa Barbara County. Damage in Santa Cruz consisted of toppled chimneys, cracked plaster, broken windows on Pacific Avenue, and structural weakening of brick buildings. To the north, the city water main was broken at Laguna Creek, and at Davenport groceries were thrown from the shelves.

These two moderately large earthquakes took place on a branch of the San Andreas system, now recognized as the San Gregorio–Hosgri fault zone. This rift splits off the San Andreas north of San Francisco near

FIGURE 2.6. Destruction of the Hotel California in Santa Barbara, 1925. *Courtesy of Santa Barbara Public Library, Edson Smith Photo Collection, public domain.*

Bolinas and has been mapped along the coastline all the way to Diablo Canyon in San Luis Obispo County. The 1926 tremors and a number of smaller shocks indicate that while it is considerably less active than the San Andreas Fault, the San Gregorio–Hosgri fault zone is active and, based on its total length, capable of generating earthquakes up to perhaps 7.5 magnitude, something to be concerned with over the long term.

March 10, 1933—Long Beach Earthquake

It was in the early evening hours of March 10 when about nine miles of the Newport-Inglewood fault zone ruptured with a 6.4 magnitude earthquake. Shaking lasted for about 10 seconds and took a major toll on the schools in the area. Within several seconds, 120 schools in and around the Long Beach area were damaged, with 70 of those being destroyed (figure 2.7). There were 120 fatalities, although experts concluded that had the shock occurred while children were in school, the deaths would have been in the thousands. It was estimated at the time that about two-thirds of the 120 deaths were a result of people being struck by falling debris while running out of buildings.

The epicenter of the quake was three miles south of Huntington Beach, so perhaps it should have correctly been called the Huntington Beach earthquake. The Newport-Inglewood fault zone extends on land for about 46 miles from Culver City to Newport Beach, where it dives

FIGURE 2.7. Damage to the John Muir School in the 1933 Long Beach earthquake. *By W. L. Huber, courtesy of US Geological Survey, public domain.*

off into the Pacific Ocean. Because of the length of the fault, it is believed capable of producing an earthquake as large as 7.4 magnitude.

Within a month of the earthquake, the state legislature passed the Field Act, which authorized the Division of the State Architect to review and approve all public school plans and specifications and also to provide supervision of school construction work. This action was in response to the recognition of the contribution of weak foundation conditions, shoddy workmanship, and substandard materials that played a major role in school failure. Most of the schools that had major failure were constructed of unreinforced masonry (brick and mortar), similar to the collapsed buildings in the Santa Barbara earthquake eight years earlier. Unfortunately, school districts across the state were allowed 40 years to bring their existing schools up to modern standards. Since 1940, no building constructed under the Field Act has either partially or completely collapsed, and no students have died or been injured in a Field Act–compliant building.

May 18, 1940—El Centro Earthquake

The 6.9 magnitude El Centro earthquake occurred at 9:35 p.m. on May 18, 1940, in the Imperial Valley, quite close to the border with Mexico, when a rupture occurred along the Imperial Fault. The Imperial

Valley–Salton Trough area is part of a complex plate boundary between the North American Plate to the east and the Pacific Plate to the west. This region is the transition zone between the East Pacific Rise spreading centers that have opened up the Gulf of California, and the continental strike-slip or transform fault of the San Andreas Fault system in California (see figure 1.4).

This earthquake was the strongest to hit this region in historic time, and while this was primarily a rural agricultural area and sparsely populated, it caused widespread damage to canals and irrigation systems and also led to nine deaths. Surface rupture was mapped over a distance of 25–37 miles with a maximum measured horizontal displacement, or offset, of 15 feet.

July 21, 1952—Bakersfield–Kern County

A few minutes before five o'clock in the morning on July 21, 1952, a 7.3 magnitude earthquake ruptured the White Wolf Fault near Bakersfield at the southern end of the San Joaquin Valley. This area is geologically complex, as several different faults converge in this portion of the state. This was also the strongest shock to hit California since the great 1906 San Francisco event. The earthquake was somewhat unique in being followed over the next two months by 188 aftershocks of magnitude 4.0 or greater and 10 of magnitude 5.0 or larger, which continued to impact already damaged structures and utilities. One idea that has been recently put forward is that the beginning of oil extraction from the Wheeler Ridge oil field three months earlier may have accelerated the initiation of earthquakes, through a process now widely recognized globally as induced seismicity.

Downtown Bakersfield was heavily hit, and many damaged buildings were subsequently demolished as the town was rebuilt with newer architectural styles. The smaller town of Tehachapi, about 30 miles to the southeast, suffered the greatest losses; 11 people lost their lives, and most of the town's buildings suffered damage. Fifteen homes were destroyed in Tehachapi, and 53 more were heavily damaged. A few miles to the northwest, two tunnels used by both the Southern Pacific and Santa Fe Railroads collapsed, six more sustained damage, and over three miles of track were bent and twisted.

February 9, 1971—San Fernando

The 6.6 magnitude San Fernando or Sylmar earthquake hit the San Fernando Valley area of Southern California early in the morning on

February 9, 1971, leaving death and destruction in its wake.[18] Shaking reached a Mercalli intensity of XI, ranking it as extreme, with strong ground shaking lasting for about 12 seconds. There were 65 deaths and 2,543 injuries from the earthquake, making it at the time the third most deadly earthquake in California history. Fortunately, the ground shook early in the morning, while the highways were relatively free of traffic and while most offices, public buildings, and schools were empty; if it had happened a few hours later, the loss of life could have been staggering.

Seismic activity in the San Fernando area had been low in comparison with the rest of Southern California during recent years. In fact, throughout recorded history only one strong destructive earthquake is known to have occurred in the immediate area.

The fault responsible for the shock was not previously thought to be a threat, which led to the realization that more thorough investigation of other faults in the area was needed. One of the challenges in any developed area is that many of the natural landforms that occur along faults have been altered by grading and development or built on, so they are no longer easily recognized. Another major finding from this earthquake was that the intensity of shaking surpassed existing building code requirements and exceeded what engineers had designed for, even though most buildings in the valley had been built during the prior 20 years. Even modern so-called earthquake-resistant structures sustained heavy damage. Most of the damage occurred in the thick alluvial soils of the San Fernando Valley where seismic shaking was enhanced relative to bedrock areas.

Those school buildings in the area of strong seismic shaking that had been designed and constructed since the passage of the California Field Act, following the 1933 Long Beach earthquake, did not suffer structural damage severe enough to have been dangerous to the occupants had the schools been in session. Older school buildings, however, suffered potentially hazardous damage, and some were condemned as a result.

Major structural damage occurred at two large medical facilities in the Sylmar area, the newer Olive View Medical Center and the older Veterans Administration Hospital. The Los Angeles County–owned Olive View 880-bed hospital complex had been built before the adoption of new construction techniques put in place after the 1933 Long Beach earthquake. Damage to the Olive View Hospital included failure of ceiling tiles, telephone equipment, and elevator doors and was excessive at the basement and first-floor levels but interestingly was minor on the upper floors. Because the first floor almost collapsed, the building was left leaning to the north by almost two feet. Three of the four external concrete stair towers fell away from the main building, and roof coverings over parking areas collapsed onto ambulances.

It is often the case that legislation is enacted after a major disaster, but typically there may be a period of many years before the new construction standards are actually required or put into practice. The 1933 Field Act, while well-meaning, is a good example. In 1967 (34 years after the act was approved) the California legislature required all pre–Field Act schools to be examined for safety by 1970; if found to be unsafe, they could not be used as public schools after June 30, 1975, unless rehabilitated to meet the standards of the act. In my hometown of Santa Cruz, an elementary school built before the passage of the Field Act was allowed to be used as a school for 42 more years until June 1975. At that time it was converted to a general public use facility under the assumption that the general public could make their own decision on whether it was safe to use. Its buildings had undergone some seismic upgrades in the interim, however.

The Sylmar Veterans Administration Hospital was opened in 1926 as a tuberculosis facility but became a general hospital in the 1960s. Twenty-six of the 45 buildings were built prior to 1933, so they had not required seismic-resistant designs. They suffered the most, with four of the buildings totally collapsing, with the loss of 44 lives. The hospital was within 3.1 miles of the fault rupture, and the structural damage was a result of strong seismic shaking of the underlying foundation materials. Because of the extensive damage, the entire complex was demolished and converted to a city park in 1972.

Two dams above the San Fernando Valley floor, the Upper and Lower Van Norman Dams, which were essentially full at the time of the earthquake, were heavily damaged. The lower dam came perilously close to breaching, with the top of the earth-fill dam partially collapsing into the reservoir, leaving it just five feet above the water level. About 80,000 people living directly below the dam were evacuated because of the threat of dam failure.

In addition to the damage to the two hospitals and the dams, there was extensive damage to single-family homes, mobile homes, businesses, schools, streets, and utilities. Water lines, for example, were broken in 1,400 places. The area is served by two major aqueducts that cross the San Andreas Fault on the way from Owens Valley; if these had broken, the region's reserve water supply would have only lasted about four hours.

Modern freeways and overpasses in the Sylmar area were severely damaged by the seismic shaking. A number of overpass spans collapsed, one killing two men parked in a truck beneath it (figure 2.8). The disruption of transportation, which tens of thousands of commuters rely on, resulting from damage of this type can greatly increase the disastrous effects of a large earthquake. The major reason for the overpass failures was the underestimation of the seismic shaking forces to be expected. Prior to the 1971 quake, there had been very few instrument readings close to large earthquakes.

One key component of seismic shaking that engineers design structures to withstand is acceleration, in both the vertical and horizontal directions. This is a measure of the change in velocity of the ground and overlying structures as the seismic waves pass. Think of this as a snapping motion, which occurs in both the horizontal and vertical. Acceleration is measured relative to gravity, or g, which is 32 feet per second per second. This is the increase in the velocity of an object each second as it falls due to gravity. Prior to San Fernando, seismologists and engineers generally believed that acceleration of the ground during an earthquake would not exceed that due to gravity, or 1 g. They did not believe that anything, whether structures or vehicles, would actually leave the

FIGURE 2.8. Freeway overpass collapse due to strong seismic shaking during the 1971 San Fernando earthquake. © 1971 Kathryn D. Sullivan.

ground during severe shaking. An instrument reading and several observations verified that they were in fact wrong. One of these was at the
location of the Sylmar Kagel Canyon fire station, where a 20-ton fire
truck was lifted up like a toy, leaving a skid mark from a tire on the wall
3 feet off the ground and coming to rest 8 feet from its original position.

The freeway overpasses that collapsed had relied on their mass to
support them on the concrete piers or abutments, with little connection.
During the shaking, the overpasses failed due to the vertical acceleration and the lack of adequate connection to the supporting piers. As a
result of the San Fernando earthquake, seismic retrofitting was required
for all of the state's overpasses that were inadequately connected to
their supporting columns.

October 15, 1979—Imperial Valley Earthquake

At 4:16 in the afternoon of October 15, 1979, a 6.5 magnitude earthquake was felt over an area of about 128,000 square miles of the Imperial Valley–Salton Trough area of Southern California and northern
Baja California. Like the 1940 El Centro earthquake, the 1979 event
occurred in an area of complex faulting where the short spreading centers of the East Pacific Rise in the Gulf of California transition to the
San Andreas Fault system that then heads northwest through 750 miles
of California (see figure 1.4). The rupture was complex; parts of the
Imperial Fault, the Brawley fault zone, and a newly recognized and previously unknown break, the Rico Fault, were all involved.

The area had been heavily instrumented, so records of ground motion
during the earthquake were extensive and added considerably to our
knowledge of seismic shaking. While observations indicated that the
overall maximum intensity of shaking was VII (very strong), the largest
damaged building, the Imperial County Services Building in El Centro,
was judged to have experienced intensity IX (violent with damage great
in substantial buildings, with partial collapse). The County Services
Building, built in 1971, was a six-story reinforced concrete structure
about 18 miles from the earthquake epicenter. Because of the nature of
the building (size, structure, and location), it was outfitted with nine
strong motion sensors, which captured very useful information on the
intensity of shaking. Several types of irregular construction styles that
had been used in the building contributed to the mass and strength not
being uniformly distributed through the structure. The end shear walls
stopped below the second floor, and the first floor was designed to carry

the load of the building through square support columns (known as a soft floor). Shaking was intense due to failure at the foundation and first-floor level, leading to this nearly new building being declared a total loss and ultimately demolished.

Field surveys documented initial lateral ground displacement along the Imperial Fault of 22–24 inches. Within five months, however, the displacement had increased to 31 inches (this is known as postseismic slip). Scientists also discovered sand boils or sand volcanoes due to subsurface liquefaction near the New River. Because of the generally rural nature of the area, there were no fatalities but 91 total injuries in California and across the border in Mexico.

October 17, 1989—Loma Prieta Earthquake

While there is no one alive today who experienced the 1906 San Francisco earthquake, many of us in the Monterey Bay and greater San Francisco Bay areas have vivid memories of the October 17, 1989, shock.[19] We can remember exactly where we were, who we were with, and what we were doing. At 5:04 p.m. on that warm October evening, 15 seconds of shaking forever changed the greater San Francisco Bay/Monterey Bay region. Three people died in older downtown brick buildings in Santa Cruz as the walls and ceilings collapsed. Survivors dug vainly with bare hands at the shattered timber and bricks in futile rescue attempts at the Santa Cruz Coffee Roasting Company. A grandmother died when the brick walls of the Bake Rite Bakery in Watsonville collapsed, and two others were killed in unlikely situations; one man died in his pickup truck on Highway One as a spooked horse running along the freeway collided with his vehicle, and another was buried by a rockfall on a north county coast beach. The sand spit under the Moss Landing Marine Laboratories liquefied, as it did in 1906, splitting the buildings down the middle and completely destroying the facilities. The Monterey Peninsula and Monterey County, in general, however, suffered relatively little damage.

As evening fell, many throughout Santa Cruz County camped outside in their yards and in vacant lots as a seemingly endless series of hundreds of aftershocks continued to shake the region and its frightened residents. Initial reports indicated that most of downtown Santa Cruz had suffered major damage and would have to be torn down. Inspections followed inspections, and in the end, many of the buildings did ultimately come down. The costs of repair and seismic upgrading were determined to be greater than building new safer structures.

The 6.9 magnitude 1989 earthquake was one more in a series of disasters over the past 175 or so years that have included earthquakes, floods, and fires and that have wreaked havoc on downtown Santa Cruz. After an earthquake, the choice for reconstruction material was typically wood. After the next fire, masonry was favored. While the county courthouse (Cooper House) had been damaged in 1906, it was upgraded and remodeled in the 1970s and became a popular Pacific Garden Mall attraction. Though it did not collapse on October 17, 1989, the earthquake damage and cost of restoration was deemed too high, and it sadly fell to the wrecking ball a week after the quake. In the first month, 20 downtown buildings were demolished, and more were to follow.

Nearly every house in Santa Cruz suffered some damage. Many older chimneys either fell or were cracked at the roofline and were ultimately rebuilt. Had October 17, 1989, been a cold fall day with fireplaces and woodstoves burning, fires would no doubt have started in many homes as chimneys collapsed and caused far more damage. People may have started calling it the Loma Prieta earthquake and fire, as they did for the 1906 San Francisco event in order not to discourage rebuilding and visitors. Fifty homes were destroyed in the city, and many other older homes were damaged. Estimates of home damage in the city reached $177 million (in 2022 dollars). Downtown business came to a halt as safety fences were placed around the hazardous and condemned structures, which at one point included 32 square blocks. Within a few weeks, a decision was made to construct a number of large tents or pavilions in the city parking lots, to house many of the downtown businesses until repairs, retrofitting, or reconstruction could take place. The subsequent months were challenging for many business owners.

At the time, the Loma Prieta earthquake was one of the costliest natural disasters in US history. Even with an epicenter in a remote area in the heavily forested southern Santa Cruz Mountains, it still caused major damage and loss of life over 60 miles away in San Francisco and Oakland (see figure 2.1). In all, 67 people died, 3,757 were injured, 18,306 homes and 2,575 businesses were damaged, and total damage was over $13.3 billion (in 2022 dollars).

The Loma Prieta earthquake ruptured a 25-mile-long segment of the San Andreas Fault beneath the central Santa Cruz Mountains southeast of Highway 17. This segment had been recognized as having the greatest chance for producing a magnitude 6.5 to 7 earthquake of any fault segment in California north of the Mojave Desert. While the location and magnitude of this earthquake were not surprising to geologists and seismologists, there were no obvious short-term precursors to warn of the impending earthquake. The Pacific Plate (or Santa Cruz side of the fault) moved northwest about 6.2 feet and upward about 4.3 feet relative to the North American Plate on the other side (San Jose side of the fault) during the earthquake.

An unusual aspect of the earthquake was the absence of recognizable surface rupture along the fault itself. From observing the effects of earthquakes of this magnitude elsewhere in the world, geologists expected that fault rupture at the surface would have occurred; instead, a 3-mile-wide zone of cracking was observed along the general trend of the fault along the crest of the Santa Cruz Mountains. Opening of many of these cracks was great enough to damage houses, roads, and utilities. This type of widespread ridgetop cracking represented a previously unappreciated earthquake hazard, one that extended well beyond the usually well-defined fault trace itself. Homes with the most spectacular views, those built on narrow ridgetops, consistently suffered the greatest damage from ridgetop cracking, particularly along the Summit area southeast of Highway 17. Many newer houses suffered complete failure due to a lack of adequate connection to foundations and/or insufficient first-floor shear bracing on two-story houses, particularly those with large two-car garage openings (figure 2.9).[20]

FIGURE 2.9. Failure of the first floor of a home in the Summit area due to inadequate shear bracing for a triple-wide garage opening. © 1989 Gary Griggs.

Many of these cracks in the Summit Ridge area southeast of Highway 17 were concluded to be the result of reactivation of large ancient landslides, as had been observed in 1906.[21] An important difference, however, was the very dry conditions in October 1989, in contrast to the saturated ground in April 1906. The hillsides in 1906 turned to mud and flowed downhill for hundreds of feet, carrying tall redwoods and destroying or burying everything in their paths. In 1989, the 15 seconds of strong shaking dislodged rocks and loose material, which slid or rolled downslope, and also disrupted the old landslides, causing cracks around their edges, but produced little overall movement.

As predicted in various hazard maps, a repeat of the 1906 earthquake damage took place where seismic shaking was amplified in areas underlain by thick deposits of water-saturated, uncompacted sand and mud. Downtown Santa Cruz and Watsonville, San Francisco's Marina District, and similar settings around the margins of San Francisco Bay suffered from enhanced shaking, liquefaction, sinking, and building tilting. This was precisely what had happened in these same areas in 1906. Many older buildings had never been brought up to modern seismic codes. The lack of reinforcing steel and the deterioration of the mortar between the bricks in older masonry buildings all contributed to the damage.

Several large fires broke out in Watsonville, and thousands of the city's 30,000 residents were dislodged from their apartments and homes. Makeshift tents sprung up in front yards, vacant lots, and parks that first night, while many others slept in their cars. The earthquake destroyed 195 homes and damaged over 1,100 more, many in the poorest sections of town, where farm laborers, packing plant workers, the unemployed, and the elderly resided. In mobile home parks, 106 mobile homes were completely destroyed as they fell off their support jacks, while another 340 were damaged. The predominantly brick downtown area suffered as 32 business properties and many historic buildings were seriously damaged. Nine hundred injured people were treated at Watsonville Community Hospital in the days immediately following the shock. The city's estimated damage to homes amounted to $326 million, while businesses suffered an additional $433 million. Losses of frozen food and farm damages reached $187 million (in 2022 dollars). With the collapse of the Highway One bridge over Harkins Slough, all of the freeway traffic was directed through downtown Watsonville, which added to the problems.

Although brief references to coastal bluff failure appeared in accounts of earlier earthquakes in the region, there were very few homes that had been built close to the coastal bluffs at the time of the last large earthquake in 1906. Residents a century ago seemed to place less value on building as close as possible to the bluff edge than do today's property owners.

Cliff failure took place during the Loma Prieta earthquake along about 100 miles of coastline, from Marin to Monterey Counties. Bluff collapse and damage were greatest between Santa Cruz and La Selva Beach, however, closest to the epicenter. Shaking produced cracking of the bluff edge, which damaged cliff-top homes. As the loose debris cascaded downslope, it damaged and blocked access to homes at beach level in places like Las Olas Drive, Beach Drive, and Place de Mer. Three bluff-top homes in the Rio del Mar area and six Capitola apartments on Depot Hill were subsequently demolished as a result of earthquake damage and foundation cracking.[22]

There are not many events that are fixed in time as precisely as a big earthquake. The town clock in Santa Cruz stopped at 5:04 and stayed that way for months. The Loma Prieta ("dark mountain or hill" in Spanish, dark indeed) shock was the largest California earthquake in 37 years and the strongest on the San Andreas since the 1906 San Francisco event. Twenty-five miles of the mythic San Andreas Fault in Santa Cruz County had broken loose, from Highway 17 southeast to the Pajaro River. The Santa Cruz side and the Pacific Plate moved about six feet to the northwest and rose vertically about four feet relative to the San Jose side in those 15 long seconds. For comparison, the 1906 earthquake ruptured 230 miles of the fault, from San Juan Bautista in the south to Point Reyes in the north. The Santa Cruz Mountains segment had been locked for 83 years, slowly accumulating strain, until 5:04 p.m., October 17, when it broke loose.

Epicenters are hypothetical points placed on maps to give us some sense of where earthquakes took place. Nearly all earthquakes occur at some depth within the earth, and an epicenter is defined as the location on the Earth's surface directly above the place where the initial failure or rupture took place, which is called the hypocenter or focus.

There is considerable uncertainty in locating epicenters, however, and in some ways it's as much an art as a science. Determining the location of an epicenter depends upon analyzing the records of a number of seismographs at different locations. The job of a seismologist is to study the record and measure the gap or difference in the arrival times of different types of seismic waves, which are related to the type of subsurface rocks between the earthquake and the seismograph. We don't know the distribution of different types of rocks beneath the ground surface well in most places, so there are some assumptions involved, which often produce significant uncertainties in precisely locating an epicenter. This is the best we can do, but it is imprecise. This lack of certainty, however, doesn't discourage many curious people who have a strong desire, almost a magnetic attraction, to get to that precise place on the map with the X on it.

So, some enterprising individual(s) decided to mark the approximate epicenter of the Loma Prieta earthquake in the Forest of Nisene Marks above Aptos, which is a well-used, but somewhat out-of-the-way, state park. The sign and location were well publicized, which brought literally thousands of people into a remote section of the park. Some visitors arrived in limos and fancy clothes, having decided to drop by on the way to a dinner party. Park rangers, realizing that this attention and traffic were taking their toll on the trail to a rather remote area of the park, decided to relocate the epicenter sign to a more accessible location. The new spot was no doubt just as accurate as the original site, and visitors received just as much satisfaction in reaching the new "epicenter," their seismic pilgrimage now complete.

Although many Central California residents fancy themselves as amateur seismologists, and immediately compare their magnitude estimates after each moderate tremor in the region, most were unprepared emotionally for the aftershocks that followed the 1989 event. A magnitude 5.2 aftershock came 37 minutes after the main event, and 33 hours later, a 5.0 shock again struck terror into those who were just beginning to calm down. A dozen magnitude 4 or greater earthquakes shook the area in the first 7 hours. In the first two days, 67 aftershocks of 3.0 or larger terrorized those still in the county. It takes a while for two very large tectonic plates to come completely unstuck, and large numbers of aftershocks can be expected to accompany every big earthquake.

June 28, 1992—Landers

At 4:57 in the morning on June 28, 1992, much of Southern California was awakened by a 7.3 magnitude earthquake. Shaking lasted for two to three minutes, and while this shock was greater than the 1971 San Fernando earthquake, there were fewer fatalities and much less damage because of its location in the sparsely populated Mojave Desert. The motion was right-lateral strike-slip, like motion on the San Andreas Fault, and the maximum horizontal displacement reached 18 feet with a vertical offset of 5.9 feet. Several different faults were involved, and the surface rupture extended for about 43 miles. By any measure, this was a significant earthquake.

Although population in the area was low, damage was severe. Buildings and chimneys collapsed, roads were buckled, and there were extensive surface cracks and fissures. Although the shaking and damage injured over 400 people, there were just three deaths, two from heart attacks and one of a young boy who died under a pile of bricks from a chimney that collapsed on him.

Large earthquakes in this area aren't common, and two ideas have been put forward by geologists about why the Landers earthquake occurred where it did. The movement between the North American and Pacific Plates may be migrating eastward from the San Andreas Fault to a new alignment across the Mojave Desert and east of the Sierras. A second idea is that this and other associated earthquakes, including the 6.1 magnitude Joshua Tree quake of 1992, may be a manifestation of the propagation of rifting of the plates moving northward from the Gulf of California.

January 17, 1994—Northridge

The 6.7 magnitude Northridge earthquake struck at 4:30 a.m. on January 17, 1994, and was the largest earthquake in California to occur under a large metropolitan area since the 1933 Long Beach shock. Although the initial name, Northridge, has stuck, the epicenter was relocated to the neighboring community of Reseda within a few days. Not unlike some other large earthquakes in California, this one occurred on a previously unmapped fault. This is not surprising in an intensively developed but geologically young and active area like the San Fernando Valley. Houses and commercial buildings, schools and government buildings, as well as streets and parking lots have covered over any natural landforms that might have provided an indication to geologists of a fault's presence.

Shaking from the earthquake lasted about 10–20 seconds, and the peak ground acceleration measured was 1.82 g, the highest ever instrumentally measured in any urban area in North America.[23] Heavy seismic shaking and damage occurred up to 85 miles away, with most of the major losses in the west San Fernando Valley and in the Santa Monica, Hollywood, Simi Valley, and Santa Clarita areas. Damage was heavy in the Northridge Fashion Center and at the California State University, Northridge campus, particularly from the collapse of parking structures. Freeway overpasses and interchanges also suffered major damage, as they did in the 1971 San Fernando earthquake, which caused major traffic and commuter disruption. Fifty miles away in the Orange County city of Anaheim, the scoreboard at a stadium collapsed onto several hundred seats, but fortunately, at 4:30 in the morning, the stadium was empty.

The greatest damage and loss of life occurred in multistory wood-frame buildings (such as the three-story Northridge Meadows apartment complex). Those buildings with poorly reinforced first floors (soft stories), such as those with open first-floor parking, performed poorly. Broken gas lines led to a number of fires as houses shook off their foundations and unsecured water heaters fell over. Somewhat expected, older unreinforced masonry buildings and houses on steep slopes performed poorly, with major damage. School buildings overall fared well as a result of the standards established by the Field Act following the 1933 Long Beach quake. California State University, Northridge, was an exception, however, with all 58 campus buildings sustaining significant damage. The fine arts building and the South Library suffered internal structural damage and were subsequently demolished and replaced. A newly completed parking structure completely collapsed and had to be demolished. Fire damage was also widespread in campus buildings. While the Field Act applied to K–12 schools, these university building losses raise some serious questions about the building codes for these educational structures. Fortunately the buildings were unoccupied at this early hour, or fatalities would likely have been much higher.

Eleven hospitals and medical facilities suffered structural damage such that a number of them were declared unusable. The state legislature subsequently passed legislation requiring all hospitals in the state to have their acute care and emergency rooms in earthquake-resistant buildings by January 1, 2005. Most of the hospitals couldn't meet this date, however, and it took several more years for them to become compliant. A 2019 estimate of the total damage from the quake reached $38 billion, making it one of the nation's most expensive natural disasters.

The total death toll was believed to be about 60, with 16 of those fatalities occurring due to the collapse of the Northridge Meadows apartment complex. Over 8,700 injuries were reported, and 1,600 people needed to be hospitalized.

SOME FINAL THOUGHTS ON EARTHQUAKES

You might have concluded from this chapter that there is both good news and bad news about the earthquake risk in California. From the previous description of damage, injuries, and fatalities from historic earthquakes in the state, it is clear that we live in a geologically active area and that we can expect earthquakes to continue to occur, lots of small ones and the occasional big one. Unlike some more localized natural hazards, there is really nothing we can do to prevent earthquakes from occurring; we can, however, take a number of steps to reduce the future damage and fatalities.

California has adopted a number of measures to reduce earthquake losses; one of the most important was the Field Act, passed after the 1933 Long Beach earthquake, that required more stringent building standards for schools. Following the Northridge earthquake, building codes for hospitals were upgraded as well. The Uniform Building Code (UBC) has been strengthened over the past several decades as measurement of the severity of shaking during larger earthquakes has improved. Connections between foundations and framing and shear wall requirements are now far stricter than 50 years ago when many older homes and other buildings were constructed. We have witnessed the types of failures that took place in San Fernando/Sylmar in 1971, Loma Prieta in 1989, and Northridge in 1994, and California has addressed those by updating building codes for all new construction. What typically have failed in more recent quakes are those structures that predate the newer standards.

It should be very reassuring to know that no one, not a single person, died in a single-family wood-frame structure during the 6.9 magnitude 1989 Loma Prieta earthquake, even though thousands of homes were destroyed or damaged. In most of the rest of the seismic world, however, whether China, Nepal, Pakistan, Iran, Turkey, Peru, or Guatemala, most of the people still live and work in unreinforced or poorly reinforced adobe, brick, or masonry buildings, which simply aren't able to stand up to severe seismic shaking. They don't have enough lumber to build with, so they use what they have, and it hasn't worked very well.

Average annual worldwide death tolls from earthquakes in the past century have been about 50,000. In the United States, however, the average annual death toll has been about 35, the great majority of these from the 1906 San Francisco earthquake (and fire). Information of this sort has led to the development of two general schools of seismic safety in the United States, within which many professionals fall: the "bathtub school" and the "what if I'm right school." The former consists of some renegade geologists and statisticians who state, with statistics like those quoted above, that you have a greater risk of dying in your bathtub than you do in an earthquake in this country. The latter group consists of the civil defense specialists and many geologists and building officials who feel we should go to the effort to be prepared for the infrequent but inevitable future seismic events. And they both have good arguments.

It may reduce your seismic stress level somewhat to know that your chance of dying in many other everyday activities is far higher than your chance of dying in an earthquake. The odds of dying from an earthquake in this country are actually lower than those from lightning strikes, dog bites, and insect stings. Odds of dying from drowning are 100 times greater, and from a motor vehicle accident about 1,000 times greater. Cell phone distractions, whether talking or texting, cause a reported 2,600 deaths and 330,000 injuries in the United States annually. Now that's a hazard we ought to be concerned about. There is no such thing as multitasking while driving an automobile.

While the odds of dying in an earthquake in this country are very small relative to almost any other risk, the property damage from earthquakes can be astronomical. The Loma Prieta earthquake produced about $13.2 billion in damage, but losses during the slightly smaller, but more urban 6.7 magnitude Northridge earthquake of 1994 were nearly $40 billion (in 2022 dollars), making it the most costly natural disaster in US history at that time, and this wasn't even a large earthquake. For much of California, the worst may be yet to come.

3

Tsunamis

Tsunami! This word commonly evokes an instant visceral response for many of us, much like *earthquake, shark,* or *mountain lion:* events that can be devastating and deadly for coastal communities or people caught in their path. The massive 2011 Japan earthquake and tsunami that produced death and destruction, and the 2004 earthquake and tsunami in the Indian Ocean that killed a reported 235,000 people were both tragic reminders of what can happen when waves 30 feet high wash quickly and often unexpectedly over densely populated low-lying coastlines.

On a personal note, while I've been 100 percent present and had the opportunity to experience and observe the effects of most of the other types of natural disasters that affect California, I've unfortunately missed out on both volcanic eruptions and tsunamis. The former is entirely understandable, with the last eruption in the state occurring in 1915, well before I was born. The 2011 Japan tsunami, on the other hand, came right up to our doorstep here in Santa Cruz. The small craft harbor suffered major damage, but as luck would have it, I was completely unaware of the tsunami arrival as I was driving to the state capitol in Sacramento for an ocean meeting the day it arrived. I hadn't been listening to the radio that morning or I might have turned around to observe the tsunami arrival in person. As luck would have it, I also missed a second opportunity on January 15, 2022, when a surprise tsunami from an explosive eruption in the Tongan islands in the South Pacific again flooded Santa Cruz Harbor. Perhaps I will get another firsthand viewing opportunity.

Just as you can make little ripples by blowing on a cup of coffee or create small waves by jumping into a pond, lake, or swimming pool, any large disturbance or sudden movement under or in the ocean can also generate waves. Enormous oscillations of water caused by large seafloor earthquakes, underwater landslides, or exploding undersea volcanoes result in waves known as seismic sea waves, or tsunamis. These waves have frequently been called tidal waves, but they don't actually have anything to do with our tides. The word *tsunami* is derived from two Japanese words: *tsu*, which means harbor, and *nami*, which means wave. Some of the greatest historical impacts of tsunamis have occurred in harbors or port cities where the wave energy has been focused and concentrated as it reaches the shoreline. Hilo, Hawaii, is a good example. Situated at the upper end of a narrow bay on the big island of Hawaii in the middle of the Pacific, it's been repeatedly damaged by tsunamis from distant earthquakes around the margins of the Pacific Ocean.

Tsunamis and Subduction Zones

Nearly all of the world's most damaging historical tsunamis have originated at trenches, in subduction zones. The Earth's largest earthquakes occur at these locations where thin, dense oceanic plates collide with and descend beneath thicker and lower-density continental plates. All but a few of the planet's subduction zones occur in a curve around the margins of the Pacific Ocean—which has been called the Ring of Fire—and this is where most of the world's very large earthquakes, volcanoes, and damaging tsunamis have originated. Moving clockwise around the Pacific basin, we have the Middle America, Peru-Chile, Kermadec, Tonga, Bougainville, Java, Marianas, Philippine, Ryukyu, Izu-Ogasawara, Japan, Kurile, and Aleutian Trenches (figure 3.1). Then closer to California are the Aleutian Trench and the Cascadia Subduction Zone, which somewhat strangely displays no seafloor trench.

Everything about tsunamis can be described in superlatives. In the open ocean, these waves will typically travel at speeds of 450 or 500 miles per hour, about the same velocity as a commercial airliner. The wavelength of a tsunami, or the distance between two crests, is typically 90 to 100 miles; this contrasts with the wavelength of the typical wind wave that breaks along the California coast—from several hundred to

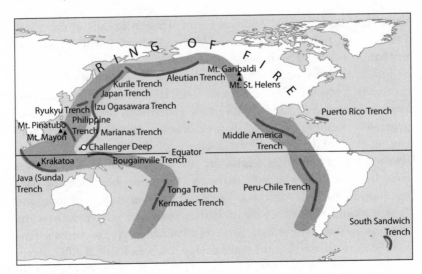

FIGURE 3.1. Map of the Pacific Ocean basin showing the location of the deep-sea trenches. *Courtesy of Gringer, public domain, via Wikimedia Commons.*

perhaps a thousand feet. With heights in the ocean of only several feet, however, and because of these very long wavelengths, tsunamis are essentially imperceptible to ships at sea and of no real concern. In fact, ships wouldn't even recognize or notice a tsunami while far from shore.

As these waves approach the coastline and move into shallow water, however, their speed and the distance between wave crests decrease at the same time that their heights increase. These long waves still have so much mass and momentum that they can wash inland considerable distances and also rise to significant elevations. During the great 2004 Indian Ocean earthquake and tsunami, waves washed inland for over two miles, and the water from at least one of the three waves reached an elevation of 100 feet above sea level. The 2011 tsunami in Japan was even more impressive as it flooded low areas up to six miles from the shoreline and reached elevations as high as 135 feet above sea level. These waves weren't actually over 100 feet high, but because of their very large momentum, they were able to push water a considerable distance up the shoreline along low-elevation coasts and reach elevations of over 100 feet in that way.

Tsunamis can approach the California coast from any of the trenches or subduction zones that nearly encircle the Pacific Ocean, from the Aleutians south to Chile, and from New Zealand north to Kamchatka (figure 3.1). Although these long-period waves might have traveled thousands of miles across the ocean from their source, because of the energy they carry, they can still cause significant damage and loss of life.

TSUNAMI HISTORY OF CALIFORNIA

Over the past 200 or so years of historical records in California, there have been 13 significant tsunamis where some damage and/or fatalities were documented. Five of these were categorized as local tsunamis where the source in each case was believed to be an offshore underwater landslide, likely triggered by an earthquake. Eight of the significant tsunamis have come from distant sources, from one of the subduction zones around the Pacific basin. Over this entire approximately 200-year period, only 17 lives have been lost due to tsunamis along the state's coastline, far fewer than the number of Californians killed by dog bites (36 fatalities in 2018 alone) and almost any other accidental cause.[1] While there have been many more small tsunamis measured by water-level recorders or tide gauges during the state's recorded history, these were typically less than a foot or two in height, caused no significant damage, and so are not included in this history. In addition, many earlier records referred to unusually large waves breaking on the shoreline as tidal waves, when in fact they were simply storm-induced or wind-driven waves, often arriving at times of high tides, thereby washing higher on the beach.

December 21, 1812—Santa Barbara Coast

Two large earthquakes on this date reportedly destroyed the mission at Lompoc and damaged the missions at Santa Barbara and Ventura. The subsequent tsunami was the first to be reported in California, although the accounts have been distorted and exaggerated over the subsequent two centuries. The consensus today is that the event was likely triggered by a submarine landslide initiated by an earthquake(s). There is now good evidence, from offshore high-resolution bathymetric surveys conducted by the US Geological Survey in recent years, for the presence of large submarine slides that are capable of generating tsunamis offshore of the Santa Barbara coast (figure 3.2).

There are a number of descriptions from this time of the ocean rising and flooding the low-lying areas along roughly 55 miles of coastline. A chapel in Ventura, located at the southwest corner of Palm and Meta

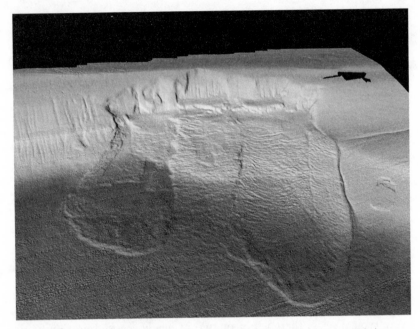

FIGURE 3.2. Multibeam bathymetric image of large slumps on the continental slope just west of Santa Barbara. *By Gary Green © Monterey Bay Aquarium Research Institute (MBARI).*

Streets, was damaged by the waves, and the site has been described as 15 feet above sea level. In Santa Barbara,

> the sea was observed to recede from the shore during the continuance of the shocks [earthquake aftershocks], and left the harbor dry for a considerable distance, when it returned in five or six heavy rollers, which overflowed the plain on which Santa Barbara is built. The sea, on its return flowed inland little more than half a mile, and reached the lower part of town, doing but trifling damage, destroying three small adobe buildings.[2]

The Santa Barbara mission was damaged by the earthquake, but it sits at an elevation of a little over 300 feet and two miles from the shoreline and was not impacted by the tsunami. Several different historic accounts suggest that the amplitude of the tsunami was about 6.5 feet at Santa Barbara and Ventura. The original Mission La Purísima was built in 1787, at what today is the town of Lompoc. It was destroyed by the 1812 earthquake and subsequently rebuilt several miles to the northeast.

November 22, 1878—San Luis Obispo County Coast

The *San Luis Obispo Tribune* reported:

> A tidal wave swept along this coast doing considerable damage to many of the landings . . . the principal damage was done at Point Sal. About half the wharf at this point is reported to have been carried away, involving the loss of several hundred sacks of grain and the drowning of one man. The Point Sal wharf was a strong structure and in thorough repair. . . . The greater part of the old People's Wharf at Avila was carried away. . . . At Morro the sea ran so high as to break over the sand ridge which divides the bay from the ocean. The Cayucos wharf was slightly damaged, losing about thirty piles.[3]

The lack of a significant earthquake at this time led to the conclusion that the tsunami was likely the result of a local submarine landslide. This was also the first reported loss of life along the state's coastline from a tsunami.

June 15, 1896—Japan Earthquake and Tsunami

What has been called the great Sanriku earthquake in Japan (magnitude 7.6) took the lives of over 26,000 people, destroyed over 11,000 homes in Japan, and produced a tsunami that arrived on the west coast of the United States. The *San Francisco Chronicle* of June 16 reported that a 9.5-foot wave arriving at low tide overtopped a temporary dike of sandbags protecting an area of the San Lorenzo River in Santa Cruz that was being used to build floats for the Venetian Water Carnival. No mention was made of damage to ships, however. The *San Diego Union* reported that a five-foot wave in Santa Cruz destroyed a protective dike and that water rose far upriver and also did severe damage to a ship moored at the pier. The *Santa Cruz Sentinel,* somewhat oddly, however, had no report of any damage from this event. The *Mendocino Beacon* wrote that "the sea rose and fell some seven feet beyond its level in mighty waves gradually becoming less."[4]

August 31, 1930—Santa Monica

A magnitude 5.2 earthquake in Santa Monica Bay occurred on a Sunday in the middle of summer when beach attendance was high, producing significant waves along the shoreline. The *Los Angeles Times* reported: "Terrible Waves, Riptide at Santa Monica Perils 16 Sunday Swimmers—Waves twenty feet high and the worst rip-tide reported on the Santa Monica beaches for years endangered the lives of the great holiday crowds yesterday and caused strict supervision of the Labor Day crowd to avoid loss of life. . . . Huge waves started at 11:30 AM believed to be caused by a tremendous disturbance of the ocean floor." The *Venice Evening Vanguard* reported: "Bathing was not at its best yesterday because of an unusually heavy sea . . . which at times reached a height of nearly 20 feet. No drowning or near drowning reported." And the *San Diego Pilot* reported that at Santa Monica "mountainous waves believed due to the earth tremor caused the death of one man near here Sunday and scores were rescued from the surf by lifeguards. Seismologists reported the temblor centered in the Pacific Ocean about 15 miles off Santa Monica and held it responsible for the heavy seas. Larry Tobin, a 20-year-old dishwasher of Palos Verdes, California, was swept beyond his depth south of Redondo Beach and was drowned."[5]

The waves were apparently confined to a 16-mile stretch of the coast from Santa Monica to Redondo Beach. Subsequent studies concluded that the waves were probably the result of an earthquake-generated submarine landslide.

April 1, 1946—Aleutian Trench, Alaska

The two tsunamis that have been the most damaging to the California coast historically were both generated by very large subduction zone earthquakes in the Aleutian Trench off Alaska—they might be called the holiday earthquakes. The first took place on April Fools' Day in 1946 and was very large, magnitude 8.6; the other, an even larger 9.2 magnitude event, occurred on Good Friday in 1964. The 1946 tsunami reached elevations of nearly 100 feet on Unimak Island, Alaska, destroying Scotch Cap Lighthouse and taking the lives of five coastguardsmen. This was one of the most important tsunamis in recent time, as it resulted in the formation of the Pacific Tsunami Warning System.

The 1946 tsunami had modest impacts in Noyo Harbor at Fort Bragg in Mendocino County on the north coast, where many boats broke from moorings and were carried upstream while others were carried offshore but were chased down and towed back into the harbor.

Princeton–Half Moon Bay was hit the most severely of any of California's coastal towns by the 1946 Aleutian tsunami. A series of waves pushed boats nearly 1,000 feet inland, flooded low-lying homes as far as a quarter mile from the beach, and washed automobiles as far as 60 feet from their parking places. At Princeton, waves washed completely over and damaged the piers and piled up logs, debris, rocks, and crates along the shoreline.

In Santa Cruz, two waves were reported, the first arriving at 10:15 a.m. and the second at 11:51, with maximum heights later documented at about 10 feet. The waves reportedly pushed water a considerable distance up the San Lorenzo River. On the municipal wharf, "lines and buoys used to fasten fishing boats suddenly went taut and ladders down the pier into the water were lifted to a vertical position as the swell passed."[6] Malio Stagnaro, of the Stagnaro Fishing Company, reported that strong currents continued to agitate the area around the wharf the following day, preventing boat owners from anchoring their boats.

There was apparently only a single tsunami fatality along the entire coast of California from that large April Fools' Day tsunami and it was in Santa Cruz. It was also the only tsunami-related death that has ever been reported in the Monterey Bay area. For some historical perspective, in 1946 tsunamis were still called tidal waves, and we had no plate tectonics to make sense of why and where these big earthquakes and tsunamis occurred. Additionally, we had no Pacific Tsunami Warning System as we do today.

About 10:15 a.m. on April 1, 1946, Hugh Patrick, a 73-year-old man walking along the West Cliff Drive shoreline near Lighthouse Point was drowned when the water level rose quickly to 10 feet above normal as the first wave hit. His walking companion, 73-year-old Cephus Smith, described in the *Santa Cruz Sentinel* as "a local dishwasher," was also knocked over by the wave and was unsuccessful in an attempt to rescue his friend. Hugh Patrick's body was recovered in the kelp beds a half mile west of Lighthouse Point 17 days later. Another man was swimming when a surge dashed him against the rocks, but he managed to fight his way out. Men on the municipal wharf reported the water receding at a terrific pace a little after 10:00 a.m. and suddenly returning at an appalling speed and surging high on the beach. There were four surges, the last at 11:50 a.m., which all but topped the esplanade along Main Beach. The *Watsonville Register-Pajaronian* reported that the wave covered the entire beach at Santa Cruz with 10 feet of water. One observer reported that the sea rose 10 feet above the normal level for the entire length of the wharf.[7]

Some fishermen in Monterey reported a slight turbulence in local waters, but little else was noticed. In neighboring Pacific Grove, a single surge of water was observed at the municipal swimming pool on the coast, which flooded the dressing room to a depth of 3 feet. One man on the steps fled ahead of the water, and photographic evidence indicates that the water reached an elevation of 10.3 feet above MLLW at Pacific Grove (MLLW, or mean lower low water, is the reference level for all tide tables, so this height would have been about 3.3 feet above the highest tides expected in Monterey).

November 4, 1952—Kamchatka

A magnitude 8.2 earthquake off the east coast of Kamchatka (the Kurile-Kamchatka Trench and subduction zone in the far northwestern Pacific—see figure 3.1) generated a tsunami that was documented along the entire coast of California, although damage was generally minor. At Crescent City, which has historically been more tsunami prone than any other coastal city in the state, there were four strong surges in and out of the harbor beginning at 8:00 in the evening. Boats were raised as high as their mooring lines allowed, but the rapid currents caught the keels of four boats that were turned over and then sank. Eyewitnesses noted that the receding water went out far beyond any low tide ever previously observed and then surged in to within four inches of the top of the wharf. Huge, 60-ton concrete mooring buoys were dragged across the floor of the harbor, snarling anchor lines.

At Santa Cruz, one of the Stagnaro Company's fishing boats (the *Bruno Madre*) was damaged when a wave struck while it was being hoisted onto the wharf. Waves rolled high onto Cowell's Beach, removing sand and leaving the lower end of Saunder's steps 10 feet above the beach (these were access stairs to the beach upcoast of the Dream Inn). Water washed up to the scaffolding along the front of the casino, whose interior was being remodeled.

In San Pedro, the local newspaper reported that water surged up and down the harbor channel and that the ferry would alternately dock too high or too low for passengers to easily disembark.

May 22, 1960—Chile

The 1960 Chile earthquake (9.5 magnitude) was the largest shock recorded during the entire past century and generated a tsunami that significantly impacted virtually the entire coast of California. In the far north, water levels in Crescent City reportedly reached 7.4 feet above predicted tide and surged into the southeastern part of the city, flooding a number of streets and leaving tons of logs and debris along Front

Street. A portion of Highway 101 entering the city was also underwater. The heaviest damage occurred at the Citizens Dock, which was covered with debris. A steel-pile retaining wall at the dock parking lot partially failed due to 6 to 7 feet of scour at its seaward toe. Three commercial fishing boats were sunk, there was flooding, and some damage was done to dock facilities. The Dock Café was flooded, and the Sea Scouts' building was floated off its foundation. Depth soundings indicated about 12 feet of sediment was deposited in some areas of the harbor.

At Noyo Harbor in Fort Bragg, surges raised water levels 4 to 7 feet above normal tide, leading to loosened or broken pilings at nearly every dock. A group of six boats broke from their moorings from the south side of the harbor and were carried upriver, where two ended up stranded on a mudbank. Waves surged up 10 feet in elevation to the pilings of the homes at Stinson Beach.

At Princeton near Half Moon Bay, the newspaper reported, the northwest corner of the bay was drained nearly dry three times, and the ocean floor was uncovered for about 600 feet beyond the main fishing pier. The falling water left a dozen or more commercial fishing and pleasure boats lying on their sides. Two salmon trawlers were carried onshore, and three men on board had to swim for their lives when waves hit one of the boats and keeled it over.

The *Santa Cruz Sentinel* reported no damage along the Santa Cruz County coast from the "tidal wave," but water washed up the steps of the Boardwalk Casino at 10:35 in the morning and also crashed over the seawall at Capitola. Six-foot waves were reported by an observer on the municipal wharf, arriving at 20-minute intervals through the morning. At Moss Landing, five-foot waves were observed every 20–25 minutes, along with severe currents in the harbor's entrance channel. The *Monterey Peninsula Herald* reported "waves surging into the bay" that completely submerged the partially completed seawall, but there was no visible damage. Water also rose to within a few feet of the city beach parking lot.[8]

Farther south at Santa Barbara, the *News Press* reported that "about 20 boats were torn loose from their moorings and the cables and chains of dozens of others were tangled. The highest swell washed into the harbor at 9:30 A.M. and rose to a height of seven feet, eleven inches, and then dropped nine feet all in less than ten minutes. A second series of high surges began again at 11:00 A.M. and had apparently subsided by 1:00 P.M. Boats were still breaking loose from their moorings and were being chased and caught by the city's harbor launch and local commercial boat operators." The water rushed into, then out of, the harbor again at speeds of five miles an hour, stirring up mud and debris from the ocean floor.[9]

The *San Pedro News Pilot* reported damage at the Los Angeles Harbor of nearly $10.7 million (this and all the values in this chapter are in 2022 dollars). Eight hundred small craft were torn from their moorings, 40 were sunk and 200 damaged. Gasoline from damaged boats spilled into the harbor, causing a fire hazard. A skin diver diving off Point Fermin was reported missing, and his body was never recovered.

In San Diego Bay, the surge swept in about every 15 minutes as a rapid increase in volume and velocity turned the channel into a swift-flowing river moving in both directions. The port master estimated the water was moving at 23–28 miles per hour with a maximum rise in elevation of 7 feet. The data from the tide gauge, however, documented the maximum rise in water elevation in San Diego at just 2.3 feet. The *San Diego Union* reported that 10 boats broke their moorings and 165 feet of dock was destroyed at the Southwestern Yacht Club at Point Loma. The wave surges broke a 100-foot bait barge in half in Mission Bay's Quivera Basin, with half smashing into a fishing dock, knocking down five moorings, and breaking eight boat slips off the dock. Some 140 feet of the harbor department's operation dock with three patrol boats also tore loose during the swirling surge.[10]

1964 Good Friday Alaskan Tsunami

Four years later, the great 9.2 magnitude Good Friday earthquake in the Aleutian Trench in the Gulf of Alaska generated a tsunami that radiated out across the entire Pacific basin. This was both the largest earthquake ever recorded up to that time and the most destructive tsunami to batter California's coast in historic time. Crescent City, on the Northern California coast, was the California city hardest hit, as has historically been the case. The city was inundated by a series of waves that pushed buildings off their foundations and into other structures and swept vehicles and buildings into the ocean (figure 3.3). Wave run-up extended 800 to 2,000 feet inland in the commercial and residential areas of the city, with water depths of up to 8 feet in city streets and 20 feet along the shoreline. The most damaging waves struck the waterfront area at 1:45 a.m., drowning 12 people, demolishing 150 stores, and littering the streets with huge redwood logs from a nearby sawmill.

FIGURE 3.3. Damage in Crescent City from the 1964 Alaskan tsunami. © *1964 Orville Magoon.*

The first wave only caused minor flooding of shops and stores near the shoreline. Local residents had experienced previous tsunami false alarms so weren't all that worried about a foot of water downtown. Gary Clawson and his father owned a tavern called the Long Branch. That evening they, along with Gary's mother, his fiancée, and two employees, came downtown to look over the tavern, clean up a little, and retrieve the cash box, and then they unfortunately decided to have another drink to celebrate Gary's father's birthday. Gary recalls, "My dad, I'll never forget, jumped on the bar, grabbed a beer and said, 'Well happy birthday to me.' And then said, 'Let it come.'" As soon as he said this, rising waters ripped the Long Branch off its foundation. After the group escaped to the floating roof, Gary Clawson found a boat to rescue his family and friends. Sadly, the boat capsized in the rough water, and everyone except 27-year-old Gary Clawson drowned. Years later, he was still trying to make sense of what happened that night, as he was spared but the waters killed his parents and fiancée. "I've lived [the 1964 tsunami] two or three times a week since it happened."[11]

Not far from the tavern, a mother tried to flee with her two children. Sadly, her 10-month-old son was swept from her arms and her 3-year-old daughter slipped and was washed away. Another woman drowned after being trapped in her car. Three others died—although no one was sure how—bringing the official death toll from the 1964 tsunami to 11, which was two-thirds of all of California's historic tsunami fatalities.

Peggy Coons and her husband Roxey were the keepers of the Battery Point Lighthouse in Crescent City at the time. They vividly recorded their observations of the fourth, and largest, wave:

> The water withdrew as if someone had pulled the plug. It receded a distance of three-quarters of a mile from the shore. We were looking down, as though from a high mountain, into a black abyss. It was a mystic labyrinth of caves, canyons, basins and pits, undreamed of in the wildest of fantasies.
>
> The basin was sucked dry. At Citizen's Dock, the large lumber barge was sucked down to the ocean bottom. In the distance, a black wall of water was rapidly building up, evidenced by a flash of white as the edge of the boiling and seething seawater reflected the moonlight.
>
> Then the mammoth wall of water came barreling towards us. It was a terrifying mass, stretching up from the ocean floor and looking much higher than the island [Battery Point, where the lighthouse was]. Roxey shouted, "Let's head for the tower!"—but it was too late. "Look out!" he yelled and we both ducked as water struck, split and swirled over both sides of the island. It struck with such force and speed that we felt like we were being carried along with the ocean. It took several minutes before we realized that the island hadn't moved. . . .
>
> Big bundles of lumber were tossed around like matchsticks into the air, while others just floated gracefully away. . . . When the tsunami assaulted the shore, it was like a violent explosion. A thunderous roar mingled with all the confusion. Everywhere we looked, buildings, cars, lumber and boats shifted around like crazy. The whole beachfront moved, changing before our very eyes. . . . The tide turned, sucking everything back with it. Cars and buildings were now moving seaward again. . . . The rest of the night, the water and debris kept surging in and out of the harbor.[12]

Most of the city's downtown was either damaged or totally destroyed by the successive waves, and rather than being rebuilt, the blocks nearest the harbor were subsequently made into a park (figure 3.4). According to the US Army Corps of Engineers, property losses approached $263 million.

Fifteen miles south, two Air Force sergeants were fishing at the mouth of the Klamath River when a wall of water about 12 feet high crashed over the sandbar at about 11:30 p.m. The men and the surrounding driftwood were picked up and carried about a half mile upriver. The men climbed onto a large log, and a second surge carried them farther up the river. They tried to swim for the north shore, but one of the two drowned on the way.

FIGURE 3.4. The portion of Crescent City that was flooded and heavily damaged during the 1964 tsunami and was not rebuilt but preserved in open space as a park. © *2019 Kenneth and Gabrielle Adelman, California Coastal Records Project, www. californiacoastline.org.*

While the Eureka Boat Basin at Humboldt Bay was little damaged, the surge did top the 10-foot-high seawall and continued into the street. The bay was filled with logs and debris, and half of the channel markers were moved off their original locations by the waves. Maximum run-up elevations reached about 5 feet at the municipal marina.

Farther south, damages from four major waves of the tsunami cost several million dollars at Noyo Harbor in Mendocino County, where 100 boats were damaged and 10–20 were sunk. Losses reached $67 million in San Francisco Bay, where docks and boats suffered considerable damage, especially in San Rafael where 4 or 5 boats were reportedly sunk. Maximum elevations of the water reached between 3.5 and 4.5 feet around the bay's shoreline.

In Half Moon Bay, Avila, and Morro Bay, boats broke loose, were damaged, and sank. Boats and harbor facilities were also damaged in Santa Monica and Los Angeles harbors, where total damages reached over $19 million.

There is a repeated pattern here: nearly all of the significant damage has been in or adjacent to ports or harbors. Under the right offshore bathymetric conditions, wave refraction can concentrate wave energy along the shoreline, which appears to be the reason why Crescent City has been ground zero and has historically suffered the most extensive tsunami damage along the entire California coast. A close look at the offshore depth contours reveals a bulge on the seafloor about 2 miles south of the harbor, which acts as a lens to focus wave energy and may well be an important reason for the city's repeated tsunami damage.

The 1964 tsunami surge raised water levels about 10 feet at the Santa Cruz Small Craft Harbor, which was just being completed. As the water receded, the harbor was drained and boats were left resting on the bottom. The 35-foot dredge that had been brought in to remove sand from the new harbor was carried by the first surge out into the bay, where it sank. Several days later, skin divers Jack O'Neill and Robert Judd located the sunken dredge 70 feet off the end of the east jetty in about 8 feet of water. The sunken vessel was pulled ashore by a Granite Construction Company bulldozer. A 38-foot fishing boat, the *Big Boy,* was damaged as it exited the harbor, perhaps hitting the submerged dredge, and quickly sank. Two men jumped overboard and were rescued by another boat. Total damage to boats and infrastructure at the new harbor exceeded $69,800,000. Water came up to the Boardwalk steps, with waves described as being 8 feet high. A 14-foot wave was reported at Capitola as water overtopped the seawall along the esplanade, which was described at the time as "a not uncommon happening at high tides."[13]

At Moss Landing, maximum wave heights of 5 feet were reported, with strong currents in the harbor entrance. Waves 8.5 feet high surged in the bay at Monterey, with waves coming in at 20-minute intervals and water elevations reaching a maximum of 7.5 feet MLLW. At Pacific Grove, maximum water elevations were measured at 7 feet above MLLW, with maximum wave heights of 6 feet.

Morro Bay was hit particularly hard by the 1964 Alaskan tsunami despite being protected by two long jetties. The fuel dock at the Morro Bay Marina broke loose and damaged several boats. The yacht club lost its houseboat, which had been moored near the south embarcadero boat launching ramp. The houseboat broke free and floated rapidly down the bay on a 20-mile-per-hour outgoing tide. It rammed into the end of a dock, completely splintered, and then proceeded to sink. An oyster barge broke loose and came down the bay with the outgoing current, destroying several pilings and running into another boat. Many small boats were broken loose from their moorings and were believed to be lost, while others ended up washed aground on a sand spit inside the harbor.

Farther south, the Los Angeles Harbor suffered damage to berths 206, 207, and 208 on the Terminal Island side of Cerritos Channel. The damage occurred when a high rapid surge entered the channel about 6:00 a.m., wrenching boats and finger piers from their moorings, with about 76–100 boats floating free. The *Santa Maria*, a Union Oil tanker, ripped out a 175-foot section of dock when it was suddenly pushed against it while being moved by tugboats.

The 1964 Alaskan tsunami was the most damaging and destructive and also produced the greatest number of fatalities of any tsunami hitting the coastline of California in the period of written historic records. Much of the damage occurred in harbors where the main losses were from the strong currents rather than the water levels or flooding. Marinas are not designed for the high-velocity flows that can occur during large but relatively infrequent tsunamis. Taking actions to move boats out of harbors to open water before the arrival of large waves, elevated water levels, and strong currents, while inherently dangerous, can be beneficial and reduce boat damages. Dealing with boats in a harbor during a tsunami can be more challenging. It would be good to consider this tsunami and its impacts in our planning for future events in terms of water elevations, areas flooded, and current velocities.

2011 Japanese Tsunami

On March 11, 2011, a magnitude 9.0 earthquake in the Tohoku region of Japan produced a moderate-height tsunami along the California coast. Although it did not generate significant flooding along the state's shoreline, strong tsunami currents caused one death and over $107 million in damages to 27 harbors statewide, with the most significant damage again occurring in Crescent City and Santa Cruz. The Pacific Tsunami Warning System was in place at the time and allowed communities along the coast of California hours of warning time prior to the arrival of the tsunami.

The earthquake, one of the largest in the last century, was accompanied by 23 to 33 feet of seafloor uplift off Japan, which generated a major tsunami that propagated both directly onshore and also offshore across the Pacific Ocean. A significant portion of the east coast of Japan was devastated as the tsunami reached maximum elevations of 133 feet above sea level, flowed as far as six miles inland, and flooded about 217 square miles of coastal land. Over 90 percent of the estimated 19,575 earthquake-related deaths were due directly to the tsunami.

The waves spread out across the Pacific and produced elevated water levels and moderate damage along the west coast of the Americas from Alaska to Chile. Between the Pacific Tsunami Warning Center in Hawaii and the US National Tsunami Warning Center, watches and warnings went out across the Pacific basin and along the West Coast from Alaska to California. With accurate advance warnings, people were notified, and evacuations of low-lying areas were carried out. As a result, there was only a single fatality along the entire 1,100-mile coast of California— a photographer standing along the Northern California shoreline photographing the incoming waves didn't take the warning seriously.

In California, damage exceeded $135 million, primarily in coastal ports and harbors (table 3.1). As has been the case with each of the state's historic tsunamis, the most severely impacted area was the Crescent City Harbor in Del Norte County. The advance warnings led to a complete evacuation of the tsunami hazard area in Crescent City and a nearly complete exodus of the harbor's fishing fleet. The peak elevation of the tsunami reached eight feet but fortunately occurred at low tide, which minimized the shoreline flooding. Strong currents and surges within the harbor continued, however, for over 24 hours. Within the small boat basin, all of the docks were either damaged or destroyed. While most boats exited the harbor prior to the arrival of the tsunami, 47 others were damaged and 16 sank. Total harbor damage, to boats and infrastructure, reached about $20 million.

Local observers noted the water receding and returning along the beaches of northern Monterey Bay. Because of the advance warnings, the boardwalk, municipal wharf, and low-lying streets near the beachfront in Santa Cruz were closed. City officials advised about 6,600 people in the potential tsunami inundation zone to evacuate. Although the warning was advisory and not mandatory, many residents in low-lying areas drove up into the hills or even as far as the crest of Highway 17. At 1,800 feet above sea level, this is playing it very safe.

The Santa Cruz Small Craft Harbor, as in past tsunamis, again received the brunt of the impact in Monterey Bay. Shortly before 8 a.m., the first of 8 to 10 surges, about 10 minutes apart, swept up the harbor. The surges were described as being more like a river than a big wave. Peak height of the waves reached 5 to 6 feet, and waves traveled up the harbor at velocities of up to 15 knots (16.5 miles/hour). Boats were slammed into each other and into some of the docks (figure 3.5). One 30-foot sailboat was partially sunk and drifted towards the harbor mouth but was stuck for a while under the Murray Street bridge. One

TABLE 3.1   2011 JAPAN TSUNAMI ELEVATIONS AND DAMAGE ALONG
CALIFORNIA COAST FROM NORTH TO SOUTH
Observations from tide gage data or interpreted by a subject matter expert

Harbors, Ports, Bays, and Docks	Arrival Time		Maximum Tsunami Amplitude (m)	Reported Damage or Other Effects from Tsunami
	Forecast	Observation		
Crescent City	7:23	7:34	2.47	Near complete destruction of small boat harbor ($20M)
Klamath River				One fatality (drowning)
Eureka / Humboldt Bay	7:22	7:34	0.97	NDR
Noyo River			0.8–1.0	Major damage to docks/boats ($4M)
Arena Cove	7:26	7:29	1.74	NDR
Bodega Bay			0.5–0.7	NDR
Point Reyes	7:39	7:46	1.35	NDR
Bolinas			0.7–0.9	NDR
Sausalito			1.2–1.5	Houseboat damage; broken sewer line
Martinez		9:50	0.06	NDR
Oakland		8:36	0.51	Minor damage at nearby Berkeley Marina
Alameda		8:36	0.51	NDR
San Francisco	8:08	8:12	0.62	Two piles broken
Pacifica			0.8–1	NDR
Half Moon Bay			0.7	NDR
Santa Cruz			1.6–1.9	Multiple docks destroyed, 14 boats sunk ($28M)
Moss Landing			2	200 piles damaged ($1.8M)
Monterey	7:44	7:48	0.7	NDR
Morro Bay		8:00	1.6	Damage to several docks and boats ($500k)
Port San Luis	8:03	8:10	2.02	NDR
Pismo Beach			0.7–1.0	NDR
Santa Barbara	8:17	8:27	1.02	Damages to barges and boats ($70k)
Ventura			1.3	Damage to dock and number of boats ($150k)
Oxnard		8:30	0.9–1.2	Minor damage to docks
Port Hueneme			1.3	Damage to dock and number of boats ($150k)

(Continued)

TABLE 3.1 *(Continued)*

Harbors, Ports, Bays, and Docks	Arrival Time		Maximum Tsunami Amplitude (m)	Reported Damage or Other Effects from Tsunami
	Forecast	Observation		
Marina Del Rey		8:30	0.9–1.0	Minor damage to docks; dinghies sunk
Redondo Beach			0.6–0.7	Docks destroyed, five boats damaged
Two Harbors / Catalina				Damage to several docks and 10 boats
Los Angeles	8:32	8:40	0.49	Minor damage to docks and boats
Long Beach				Minor damage to docks and boats
Huntington			0.72	Boat pulled off mooring
Newport		8:46	0.3	NDR
Dana Point		8:30	0.6	Pylon damages when hit by boat
Oceanside			0.5	NDR
La Jolla	8:41	8:47	0.39	NDR
Mission Bay				Dock destroyed, 13 boats damaged ($136k)
Point Loma			0.5	NDR
North Shelter Island, San Diego Bay			0.3	NDR to north Shelter Island; however, a boat sunk and there was damage to dock in south Shelter Island
Navy Pier, San Diego Bay			0.63	NDR
Marriot Marina, San Diego Bay			0.6	NDR
Imperial Beach		9:30		NDR

NDR = Not Detected nor Reported.

NOTE: Courtesy of the California Department of Conservation, public domain.

dock "just blew up. It buckled and splintered" according to one observer. When it was all over, about 30 boats had broken free from their dock moorings, 14 had sunk, and dozens of others were damaged. Of the harbor's 29 docks, 23 sustained significant damage ranging from severe cracking of floats to complete dock destruction. Total losses to harbor facilities, including docks, pilings, and other infrastructure, as well as boats, ultimately reached about $28 million. Damage and

FIGURE 3.5. The 2011 tsunami from Japan left major damage to boats and boat docks in the Santa Cruz Small Craft Harbor. *Courtesy of the Santa Cruz Port District, public domain.*

coastal flooding could have been much worse, however, as the tsunami from 5,000 miles away arrived at low tide.

In Monterey, the tide gauge showed four distinct pulses of water, reaching about 2.2 feet above normal sea level but no significant damage. At Moss Landing the tsunami surge did cause five- to eight-foot waves in the harbor, but with Elkhorn Slough directly behind the entrance channel, there was a large area where the wave energy could dissipate. There was initially little observable damage to vessels, docks, or other parts of the harbor. Subsequent engineering surveys, however, found that 200 timber pilings used to anchor docks and berthing facilities had been damaged by wave scouring during the tsunami. About $1.8 million in federal emergency funds were approved to remove the damaged pilings and replace them with precast concrete pilings.

A student at the Moss Landing Marine Laboratories made some observations during the 2011 tsunami:

> And what did happen in Moss Landing? Those that were here reported that the slough in front of the labs drained, then filled up about 3 feet higher than was normal for the tide, then rapidly drained again as the surge rushed back

out. This happened repeatedly and could also be seen in the harbor. Even though some of our facilities on the sand spit are on much lower ground than the main labs, no tsunami surges breached the land and no flooding occurred. The main lab, up on our hill, was actually used by some of the locals for high ground and refuge from possible tsunami waves.[14]

The 2021–22 Hunga Tonga-Hunga Ha'apai Eruption and Tsunami

In mid-December 2021, an eruption began on Hunga Tonga-Hunga Ha'apai, a submarine volcano in the Tongan islands in the southwestern Pacific. The Tongan islands, about 400 miles east of Fiji, are part of the Tonga-Kermadec chain of volcanic islands and lie over a subduction zone (see figure 3.1). While this eruption began calmly enough, this all changed a few weeks later on January 15, when the volcano exploded violently. This catastrophic eruption was the largest globally since the eruption of Mount Pinatubo in the Philippines in 1991 and has been reported as the most powerful since the famed 1883 eruption of Krakatoa in the East Indies. By any measure, this was a massive event, and it was to have impacts on shorelines around the entire Pacific Ocean.

The explosive eruption generated tsunamis that spread around the margins of the Pacific basin, initially impacting Tonga, Fiji, American Samoa, and Vanuatu. The waves then propagated north and west to the shorelines of Japan, Taiwan, South Korea, and the Kuril Islands, but they were all relatively low along these coasts, ranging in height from less than a foot to 4.5 feet. After traveling about 5,000 miles, the tsunami was recorded at tide gauges along the entire coast of California. The arrival coincided with the high tides, which increased the water levels and potential for shoreline flooding. In contrast, the tsunami from the 2011 Tohoku, Japan, earthquake arrived at a lower tide, so water levels and flooding were less damaging.

The highest tsunami waves along the California coast were documented at Port San Luis (8.5 feet), Arena Cove (7 feet), Point Reyes (5.7 feet), Crescent City (5.6 feet), and Monterey (4.6 feet). This was an unexpected event, and because of its arrival at high tide, shoreline flooding was extensive, with harbors at Ventura and Santa Cruz experiencing the greatest damage (figure 3.6).

The Cascadia Subduction Zone

The closest "local" source for a large tsunami that would affect the California coast is the Cascadia Subduction Zone that extends about

FIGURE 3.6. Flooding of the parking area and restrooms at the Santa Cruz Small Craft Harbor during the Hunga Tonga-Hunga Ha'apai eruption and tsunami. © *2022 Shmuel Thaler / Santa Cruz Sentinel.*

550 miles from offshore Cape Mendocino north to Vancouver Island (see figure 1.4). This subduction zone trends parallel to the coastline (as do almost all subduction zones), so most of the energy from a tsunami generated here would be transmitted at right angles to the zone. For Cascadia, most of the tsunami energy would head either west out into the North Pacific towards Japan, or east, directly onshore, where damage and impacts would be greatest and warning time shortest. Some of this wave energy would also travel north and south along the California coast, but it would be much less than what would strike the coast directly onshore.

Discoveries over the past 30 years by field geologists studying preserved sediments along the coastlines of Northern California, Oregon, and Washington have provided evidence of a large tsunami that struck this area 324 years ago. The offshore area is a boundary where one small, dense oceanic plate, the Juan de Fuca, collides with the huge North American Plate and is forced down beneath the continent (see figure 1.4). As the Juan de Fuca Plate slowly descends, there is tremendous friction to a depth of several hundred miles as it scrapes against the bottom of the North American Plate and pulls the outer edge of the plate downward. Most of the time these two plates are locked, but when the accumulated stress is great enough for the two plates to shift and uncouple, the edge of the overlying North American Plate rebounds upward and a truly massive amount of energy is released. As this occurs, the upward motion of the huge slab of seafloor displaces a large amount of ocean water, which typically produces a set of large waves or a tsunami.

There is mounting evidence along the coastline of the Pacific Northwest that very large seafloor earthquakes (magnitude 9) and resultant tsunamis occur every several hundred years in the offshore area between Cape Mendocino, in Northern California, and Puget Sound. Some of the newer discoveries suggest that these huge waves moved a considerable distance inland into bays and estuaries and left behind clean beach sands within the muddy organic material that would normally be deposited in these more protected low-energy environments. In addition, during a massive earthquake of this type, a large portion of seafloor and shoreline may suddenly move upward or downward. Large old trees that formerly lived a few feet above sea level were submerged and died when their roots came into contact with salt water when the shoreline sank centuries ago, creating what have become known as ghost forests (figure 3.7). They are still preserved along the coasts of Oregon and Washington and can often be observed at very low tides. The growth rings on these old trees can be counted, and the age of the trees can also be documented using carbon-14 dating. From a number of investigations of this sort—studying the sediments left behind in estuaries, the trees that died, turbidity current or submarine mudflow deposits on the deep-sea floor, and the ages of these materials—we have good evidence that very large earthquakes occur along the Cascadia Subduction Zone about every 300–500 years or so, on average.[15]

Scientists have also recently uncovered written records at a monastery in Japan, 4,500 miles away on the opposite side of the North Pacific, providing evidence that the last major earthquake on the Cascadia Subduction Zone generated a tsunami on January 26, 1700, that reached all the way to the coast of Japan.[16] Confirming this date and knowing that great Cascadia earthquakes and tsunamis occur about every 300–500 years has significantly increased our level of awareness and concern about when another very large earthquake of this magnitude will occur, and how the associated tsunami might impact the shoreline of California.

The California Office of Emergency Services, working with other state and federal agencies, has been developing tsunami risk or inundation maps for the populated areas of the central and northern California

FIGURE 3.7. Ghost forest exposed at low tide along the Washington coast at Willapa Bay. These trees dated to the 1700 earthquake and tsunami, when they were killed by tidal submergence when the shoreline subsided. *By Brian Atwater, courtesy of USGS, public domain.*

coastline, an effort that has also led to the posting of tsunami warning signs in the state's low-lying coastal communities. The Cascadia Subduction Zone is less than 100 miles offshore from Northern California, so for cities like Eureka and Crescent City, where the potential damage and threat to people is very high, any tsunami announcement would provide only minutes of warning time to coastal residents. Proceeding farther south, coastal communities would have progressively more warning time—30 minutes to an hour or two—than those north of Cape Mendocino.

For tsunamis generated from large earthquakes anywhere else around the Pacific Rim (the Aleutian Trench or Japan Trench, for example; see figure 3.1), our warning time is on the order of five to eight hours along the California coast. With a tsunami warning system now in place throughout the Pacific basin, residents of communities along the state's coast would be informed with ample time to make plans to evacuate very low-lying coastal areas.

Over roughly 175 years of observations, maximum elevations reached by tsunamis from large earthquakes around the Pacific basin have been about 10 feet on two different occasions (in 1946 and 1964, both from very large earthquakes in the Aleutian Trench off Alaska). The most extreme run-up conditions would occur when a tsunami arrived during a very high tide, which would lead to these waves extending farther inland, reaching higher elevations and flooding larger areas.

Possible Tsunami from a Large Landslide or Slump in Monterey Submarine Canyon

While the offshore Monterey Submarine Canyon in Monterey Bay is sometimes referred to by the media as a "trench," it is not a trench in the geological sense. It isn't a subduction zone where plates collide and large seafloor earthquakes occur; rather, it is a seafloor drainage system, where sediment is transported from the shoreline at Moss Landing into very deep water miles offshore by submarine turbidity currents. Seafloor investigations using geophysical tools have revealed a number of large slump scars along some of the steeper canyon walls, indicating that very large failures do occur from time to time. While there have been no direct observations of tsunamis initiated by such a process along the Monterey Bay shoreline, and we have no idea when such an event may occur in the future, the possibility exists of such a local tsunami, most likely during a large earthquake on one of the active faults in the area. Models developed to approximate a local tsunami that could occur, given the size of these ancient slumps, indicate that a wave 3–4.5 feet high could reach the beaches around Monterey Bay in 6–12 minutes—not a lot of time, but also not a very large wave.

SOME FINAL THOUGHTS ON TSUNAMIS

The size of a tsunami and its impact can vary widely depending on the magnitude of the earthquake and the associated seafloor displacement, the offshore bathymetry or bottom topography, the geometry of the shoreline, the coastal topography, and also the tidal stage when the tsunami reaches the shoreline. Because most of the tsunamis approaching the coast of California recorded over the past 200 years have come from either Alaskan or South American source areas, they must pass over many miles of shallow continental shelf before they reach the coastline. As a result, wave energy is significantly reduced, and damage has historically been far less than in many other areas around the Pacific basin, such as Japan in 2011.

Although destructive tsunamis are not everyday events in California, they do occur, and there is no question that they will occur in the future. To provide some perspective, 13 or 14 destructive tsunamis have reached the California coast over the past two centuries and only 17 fatalities have been reported. The tsunamis of 1946, 1964, and 2011 caused the most damage along the California coast and are good indicators of what could happen in the future from a large subduction zone earthquake from somewhere around the Pacific basin. The very small number of tsunamis that have had any significant effect on the California coast during historic time, compared to the far greater number of earthquakes and floods, for example, is likely a good indicator of the overall long-term risks posed by this hazard.

A large landslide or slump in the head of Monterey Submarine Canyon, offshore from Moss Landing, could generate a modest tsunami, although how frequently such events occur is not known, and models indicate that waves would not likely exceed five feet in height at the shoreline. Such an event could be generated by slope instability but would more likely be from a very large earthquake on the San Andreas Fault or the offshore San Gregorio Fault. It is also likely that both the 1812 and 1878 Southern California tsunamis were a result of large offshore slumps or landslides.

Relative to the other natural disasters we face and have experienced in California, the historical record indicates that the impacts of tsunamis from several very large earthquakes around the Pacific basin have been relatively minor. While there has been repeated damage to several of the state's coastal harbors from these past events, and 17 lives have been lost over the past 200 or so years, we have a tsunami warning system in place that provides ample warning time to retreat from the shoreline, although we need to respond to the warning. North of Cape Mendocino, however, particularly in the coastal cities of Eureka and Crescent City, there is a high risk of major damage, injury, and loss of life when a great Cascadia Subduction Zone earthquake does occur, which will be accompanied by a very large tsunami. How a community or city can plan for or adapt to a potentially catastrophic event like a large tsunami, although one that only occurs every several hundred years, is a challenge that hasn't been resolved or met by any of California's coastal communities to date.

In our own frequent road trips up the coast of Northern California, Oregon, and Washington, there are evenings along the route where we typically will find a bed and breakfast, inn, or motel to spend the night, in communities like Crescent City or Eureka, which are both very low-lying communities. I have to admit I don't sleep well on those evenings as I lie there knowing that I'm in the direct path of a potential tsunami from a very large Cascadia Subduction Zone earthquake. Is this going to be the time when the 324 years of accumulated strain is going to be released, or will I live to go on for many more years?

4

Volcanoes and Vulcanism

While I've had my own personal encounters with volcanoes, none of these has been in California. Getting close enough on a ship to see the steam as lava flows entered the sea in Hawaii and hiking across the temporarily quiescent crater of Kilauea Iki, climbing up the still-steaming volcano in the crater now flooded with Mediterranean Sea water in the middle of the Greek island of Santorini, and hiking down the inside of Mount Mazama and swimming in the cold water of Crater Lake—each impressed me with the magnitude and impacts of historical or future eruptions of any of these volcanoes.

About three hours after sunrise on Sunday, May 18, 1980, a stillness unlike any other descended along the shoreline of Spirit Lake in south-western Washington state. The spring air, just beginning to warm, was heavy with the scent of Douglas fir; the lake's calm, glassy surface provided a mirrorlike reflection of Mount Saint Helens, rising 6,500 feet above it to the south. Suddenly, the northern face of the mountain, swollen and distorted for nearly two months by pressure from hot magma and gases beneath it, began to collapse. Hot gas and ash, along with massive chunks of rock and ice, catapulted from the weakened bulge as a long-dormant volcano surged violently to life.

The blast that shot out of the north side of Mount Saint Helens that morning took almost everyone by surprise. Its force and ultimate impact were beyond most people's comprehension. Moving at nearly 180 miles an hour, the hot volcanic cloud tore over the low hills and valleys below the north side of the mountain, leveling and destroying anything in its 230-square-mile path. David Johnston, a US Geological Survey (USGS) vulcanologist who was at an observing station on the north side of the mountain where the growing bulge was being monitored with a laser, had just transmitted his last words—"Vancouver, Vancouver. This is it"—as the catastrophic eruption began; he and over 60 others lost their lives during the eruption that spring morning.

Mount Saint Helens became the first truly active volcano within the limits of the continental United States since Lassen Peak in Northern California erupted in 1914 and 1925. Until the 1980 Mount Saint Helens eruption, most people in the United States didn't really consider the Cascade volcanoes that stretch from Lassen Peak and Mount Shasta in the south through Oregon and Washington as significant geologic hazards (figure 4.1). The USGS had, however, been mapping and monitoring individual Cascade peaks. In a report on Mount Saint Helens published in 1978, they had concluded that there was a high likelihood of Saint Helens erupting within the next hundred years, and potentially before the end of the century. They were right on target.

In Hawaii and a number of other areas around the Earth, volcanic eruptions are more common, and we are continually learning from these events and the kinds of processes that occur and the areas that are affected. Through careful field studies and dating of past eruptions, geologists and geophysicists are unraveling the histories of individual volcanoes and developing tools and methods to allow them to predict potential future eruptions and their impacts.

The great majority of the planet's volcanoes are found in three different geologic settings. Those like the Cascades occur as chains or arcs of volcanoes located landward of ocean trenches or subduction zones, where thin, dense oceanic plates are being forced down or subducted as they collide with thicker but lower-density continental plates (figure 4.2). The chains of volcanoes resulting from this tectonic process circle nearly the entire Pacific Ocean basin and are often referred to as the Ring of Fire (see figure 3.1). You can follow this ring along the edges of South and Central America; extending from Northern California into Oregon and Washington; out along the Aleutian Chain; and then down the western side of the Pacific from Kamchatka to Japan, the Philippines, and New Zealand. There are two subduction zones with accompanying volcanoes in the Atlantic—the Puerto Rico and South Sandwich

FIGURE 4.1. Major volcanoes of the Cascade Range. *Courtesy of US Geological Survey, public domain.*

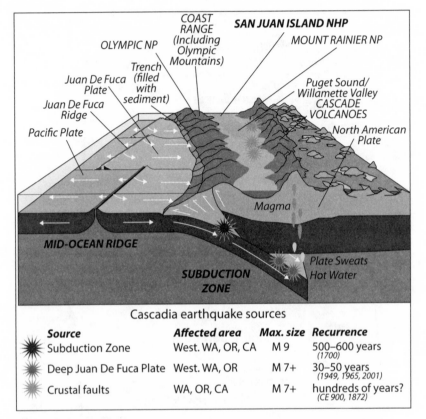

FIGURE 4.2. Cross section of the Cascadia Subduction Zone showing the origin of Cascade volcanoes. NHP = National Historic Park; NP = National Park. *Courtesy R. J. Lillie and National Park Service.*

Trenches—and in the subduction zone in the Mediterranean are the well-known volcanoes Etna, Vesuvius, Santorini, and a few others.

The approximately 50,000-mile-long oceanic ridge system is another site of extensive vulcanism. This system includes sites where oceanic plates are diverging as hot molten material rises from the Earth's mantle. These oceanic ridges and rises, however, are quite far from populated areas so don't generally represent a threat to the human population.

In the third group of volcanoes are those that occur where hot thermal plumes from the mantle erupt at the surface, whether on land or in the ocean, and build volcanoes. As tectonic plates move over these hot spots, chains of islands can form, as the Hawaiian Islands did. There are somewhere between 40 and 50 known hot spots around the planet that have formed the volcanoes of the Hawaiian Islands, Iceland, Samoa, Tahiti, the Galápagos, the Azores, and dozens of others.

VOLCANIC PROCESSES AND HAZARDS

For those of us living in California, it is only the first of these geologic settings that is perhaps of greatest concern, those volcanoes associated with the Cascadia Subduction Zone. There are, however, a few somewhat anomalous areas of potential concern in the state that will be discussed later. The Cascade Range of volcanoes extends from the Pacific Northwest into Northern California, with Mount Shasta, Medicine Lake, and Lassen Peak being the major volcanoes of interest. The Cascades are comparable in formation and shape to the other volcanoes of similar origin surrounding the Pacific basin, including the well-known peaks of Fujiyama in Japan, Arenal in Costa Rica, and Pinatubo in the Philippines.

The fundamental driving force behind the Cascade eruptions, as well as the others around the Pacific Ring of Fire, is plate tectonics. As an oceanic plate collides with the continental plate (in the case of California, this is the offshore Gorda Plate and the onshore North American Plate), the thinner and denser oceanic plate descends beneath Northern California, Oregon, and Washington. The Gorda Plate sticks to the overlying plate until the stress exceeds the frictional resistance between the plates and there is finally a break or rupture producing a very large earthquake, typically of magnitude 8 or 9. At a depth of about 200 miles, as the descending oceanic plate gets progressively warmer, it begins to dewater, or "sweat" water. The hot fluids created begin to rise and partially melt the surrounding rock on the way to the surface, where the molten rock erupts as fluid lava and forms a volcano over time.

These classic-shaped, steep-sided volcanoes are formed from viscous, or sticky, magmas that tend to produce violent explosive eruptions due to the gases contained in the molten material. These explosive eruptions can blast rock fragments and ash thousands of feet into the air. As this hot ejected material accumulates around the vent or the eruption center, a steep-sided composite volcano, or stratovolcano, forms. Mount Rainer, Mount Saint Helens, and Mount Hood in Washington and Oregon are all composite volcanoes with classic shapes.

The ejected material, called tephra, is described as pyroclastic (literally "fire-broken") and ranges from fine volcanic ash (which is actually more like fine broken glass than the ash that comes out of your woodstove or fireplace) to large fragments of rock. These violently erupting volcanoes can also produce lava flows, as well as fiery avalanches or pyroclastic flows (formerly called *nueés ardentes*), poisonous gases, and lahars (volcanic mudflows). Eruptions of different composite volcanoes, or even the same volcano over time, can vary widely depending upon the composition of the magma, its gas content, and the surrounding topography and material (snow, ash, lava flows, etc.).

Hot, flowing avalanches of volcanic ash have been responsible for many fatalities during many eruptions, most notably the complete destruction of Saint-Pierre on the West Indies island of Martinique and the death of its 30,000 inhabitants in 1902. These events consist of very hot, mobile clouds of volcanic ash and dust that are buoyed up by the hot gas entrapped and expanding within the pyroclastic material. They can flow rapidly downslope at speeds of 35 to over 75 miles per hour. Driven by gravity, fiery avalanches generally flow downhill along valleys or other topographic low areas; however, their velocities are high enough that they can flow over topographic barriers. The deposits left by these flows are known as welded tuffs or ignimbrites and can cover huge areas. Some have been mapped extending for up to 100 miles with volumes of as much as 240 cubic miles.

Seven thousand seven hundred years ago, fiery avalanches accompanied the catastrophic eruption of Mount Mazama (now the site of Crater Lake) across the border in southern Oregon. The hot, fiery clouds raced as far as 36 miles down winding valleys at temperatures of hundreds of degrees Celsius. They incinerated thick stands of timber and everything else in their path. Some canyons were filled with material up to 250 feet deep, including pumice boulders up to 6 feet across. These hot flows were able to move quickly for considerable distances, even across level ground, due to (1) the enormous volume of the flows, (2) the steepness of the descent, (3) the intense heat, and (4) the buoyant and lubricating effect of the entrapped and expanding gas.

Lahars, which are mudflows originating on the flanks of steep-sided volcanoes, have probably inundated more land and destroyed more property than any other type of volcanic hazard. They commonly occur on the steep slopes of stratovolcanoes such as Mount Shasta because of the presence of thick deposits of loose, potentially unstable pyroclastic material, including fine volcanic ash. Water is all that is needed to mobilize this mixture, and it is often available due to (1) water released from a crater lake, (2) the rapid melting of snow or ice by hot tephra, (3) heavy rains on the steep flanks of a volcano, and (4) an earthquake accompanying an eruption or explosion-induced avalanche. While these events are often induced by eruptions, they need not be, and severe rainstorms or earthquakes can trigger a lahar years after an actual eruption.

Lahars are dangerous because they typically move very rapidly (up to 60 miles/hour on steep slopes) and can be driven downhill by gravity for many miles. Slope steepness is the primary factor affecting their velocity and the distance they can flow. The material carried by a volcanic mudflow can range from fine-grained mud to blocks of rock 5 to 10 feet across. Because of their high velocity, flows can pick up and incorporate anything in their paths. Higher up on the steeper flanks of a volcano, lahars can be erosive, but as the slopes diminish, the material carried is deposited. This may be miles from the volcano, and it can even enter streams, producing a thick soup, as occurred below Mount Saint Helens in the 1980 eruption. These deposits may cover roads, railroads, towns, homes, and vehicles, and they set up much like concrete as they dry, effectively sealing off anything they have buried.

POTENTIAL VOLCANIC HAZARDS IN CALIFORNIA

In order to provide an up-to-date and easily accessible information source for potential volcanic hazards in California, the USGS established the California Volcano Observatory in 2012. This virtual observatory replaced the former Long Valley Observatory, which was originally set up in 1982 to monitor the Long Valley Caldera and Mono-Inyo Craters region, which had been showing signs of restlessness. The Volcano Observatory now monitors all of the potentially hazardous volcanic regions in California (and Nevada) in order to assist communities and government agencies to better understand, prepare for, and respond to potential volcanic activity.

The observatory has designated 11 young volcanoes within California as having moderate threat potential, high threat potential, or very high threat potential (table 4.1). Partially molten rock or magma exists below at least several of these—Medicine Lake Volcano, Mount Shasta, Lassen Volcanic Center, Clear Lake Volcanic Field, the Long Valley Volcanic Region, Coso Volcanic Field, and Salton Buttes—which produce seismic activity, toxic gas emissions, hot springs, and/or ground movement.[1]

HISTORY OF CASCADE VOLCANIC ERUPTIONS

Many years of careful mapping and dating of the various volcanic deposits around the individual Cascade peaks have enabled vulcanologists to

TABLE 4.1   POTENTIAL VOLCANIC THREATS IN CALIFORNIA

Very High Threat Potential	High Threat Potential	Moderate Threat Potential
Lassen Volcanic Caldera	Clear Lake Volcanic Field	Coso Volcanic Field
Long Valley Caldera	Medicine Lake	Mammoth Mountain
Mount Shasta	Mono-Inyo Chain	Mono Lake Volcanic Field
	Salton Buttes	Ubehebe Crater

NOTE: Based on data from "2018 Update to the U.S. Geological Survey National Volcanic Threat Assessment" by John W. Ewert, Angela K. Diefenbach, and David W. Ramsey, page 17 (USGS California Volcano Observatory).

decipher the eruption history of these mountains (figure 4.3). Although the potential hazards to life and property from volcanic eruptions were deemed minor in comparison with those from flooding and earthquakes, the threats posed by Cascade peaks were reevaluated following the damage and fatalities from the 1980 eruption of Mount Saint Helens.

In 1978 the USGS published a report entitled *Potential Hazards from Future Eruptions of Mount St. Helens Volcano, Washington*. The two authors pointed out that Mount Saint Helens had been more active and more explosive over the past 5,400 years than any other volcano in the conterminous United States (see figure 4.3). Although they didn't attempt to predict the exact timing of the next eruption, they did state that it could occur before the end of the century (before the year 2000). The nature of the volcanic products expected during a future eruption and the threats to people and property were also evaluated and delineated. The lahars, floods, and tephra (volcanic ash) from the 1980 eruption inundated and devastated areas that were almost identical to those outlined in the USGS report based on careful mapping of the deposits from prior eruptions. If anything, the warnings of potential inundation areas were conservative, in large part because of the difficulty of unraveling the geologic record and the lack of any recent historic eruptions.

Despite the geologic history and two months of clearly observable precursor activity, billions of dollars in destruction took place and 57 lives were lost from that May 18, 1980, eruption. The following facts point out some realizations that we must learn to accept and deal with in California regarding future volcanic eruptions:[2]

1. The infrequent nature of Cascade eruptions and our almost complete lack of historical and observational data for these volcanoes make it extremely difficult to predict when future eruptions will occur with any certainty.

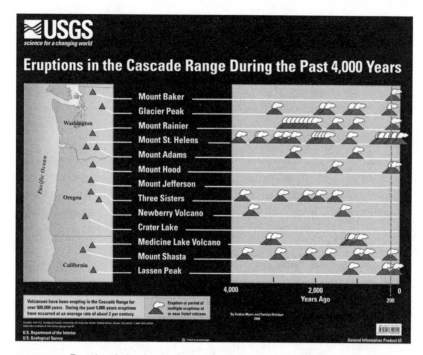

FIGURE 4.3. Eruption timing for the Cascade volcanoes over the past 4,000 years. *Courtesy of US Geological Survey, public domain.*

2. Given the current level of development around these peaks (Mount Shasta and Lassen Peak in California), including recreation areas, homes, farms, communications, utilities, and highways, and the potential for a volcano to affect areas many miles from the peak itself, an eruption in the Cascades will no doubt lead to widespread damage and, very likely, fatalities.

3. The difficulties involved with attempting to limit or halt human activities in a large areas for some indefinite period of time provide some perspective on the challenges involved in attempting to protect lives and property during future eruptions.

Lassen Peak

Prior to the 1980 eruption of Mount Saint Helens, the last Cascade eruption was that of Lassen Peak in Northern California between 1914 and 1917. The peak is the southernmost of the Cascades and rises to an elevation of 10,457 feet above the northern Sacramento Valley. Lassen is a lava dome, and with a volume of 0.6 cubic miles, it is the largest lava dome on the planet. The mountain has a complex history, originally arising from the northern flank of the now-eroded Mount Tehama about 27,000 year ago from a series of eruptions. Glaciation has eroded much of the peak since its initial formation, and many areas of the flanks are now covered with talus (figure 4.4).

Within the last 825,000 years, hundreds of explosive eruptions have come from vents scattered across an area of about 200 square miles. Around the Lassen Volcanic Center there have been over 50 nonexplosive or effusive eruptions over the past 100,000 years. The area has been relatively quiet for the past 25,000 years, however, with three notable exceptions: the Chaos Crags eruption (1,100 years ago), the eruption of Cinder Cone (in 1666), and then the eruption of Lassen Peak itself (1914–1917; see figure 1.3).

No matter how many years have passed, the time since Lassen's last eruption clearly doesn't indicate that the volcano won't erupt again. Nonetheless, much of the population living in Northern California in the early years of the last century believed that the volcano was extinct and wouldn't erupt again. What is important to keep in mind with all of the Cascade volcanoes is that they are still connected to an active subduction zone. The two tectonic plates are still colliding. The Gorda Plate partially melts as it descends into the Earth's mantle, and the resulting hot fluids rising through an internal plumbing system that periodically reaches the surface can produce a variety of eruption products. As figure 4.3 illustrates, all the Cascade peaks but two have erupted intermittently over the past 4,000 years, with some of these being more active than others. For the Northern California Cascades, our major concerns are with Medicine Lake Volcano, Mount Shasta, and Lassen Peak.

The most recent Lassen Peak eruption consisted mostly of irregular steam blasts, which began on May 30, 1914, now a little over a century ago (see figure 1.3). With no apparent precursor earthquakes, the volcano suddenly erupted with a huge volume of steam and a series of explosions, which carved out a small crater with a fairly deep lake on the volcano's summit. A year later on May 14, 1915, however, partially molten rock began flowing from a vent and traveled as far as Manton, 20 miles west of the mountain. By the next day, the volcano had built an unstable lava dome. The dome collapsed during an eruption a few days later on May 19, which sent an avalanche of hot rock down the north side of the volcano, covering an area of nearly four square miles at the base of the mountain. The ash and fragments of lava combined

FIGURE 4.4. The northeast side of Lassen Peak. *Courtesy of US Geological Survey, public domain.*

with snow on the volcano's flank initiated a high-velocity lahar, or mudflow, which flowed four miles down the east side of the mountain and filled 35 miles of the Hat Creek valley virtually overnight (figure 4.5). After being deflected to the northwest at Emigrant Pass, the lahar continued an additional 7 miles to Lost Creek. On May 20, the lower Hat Creek valley was flooded with muddy water, which damaged ranch houses in the Old Station area, removing homes from their foundations. A few people received minor injuries as the muddy floodwaters continued for another 30 miles, killing fish in the Pit River.

About four in the afternoon on May 22, 1915, the volcano generated another violent explosive eruption that ejected rock and pumice and formed a larger and deeper crater at the summit. Three days later, a vertical column of ash exploded from the vent and reached an altitude of 30,000 feet; it was visible in Eureka, 150 miles to the west. That column partially collapsed and generated a pyroclastic flow composed of hot ash, pumice, rock, and gas that covered three square miles of land and spawned a lahar that stretched 15 miles to Hat Creek valley again. Smaller mudflows formed on the other sides of the volcano, and volcanic ash was also noticed up to 280 miles to the east in Elko, Nevada.

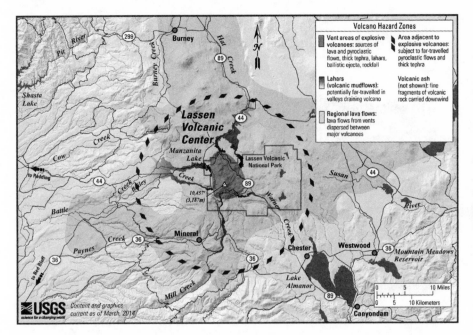

FIGURE 4.5. Lassen Volcanic Center showing the potential impact area for ground-based hazards during an eruption event. *Courtesy of US Geological Survey, public domain.*

The presence of very hot rock beneath Lassen Peak's surface was indicated by steam explosions that continued for several more years. An especially strong steam explosion in May 1917 formed the northern crater at the mountain's summit. In June, 21 additional eruptions were reported and altered the crater, creating a new vent on Lassen's north-western summit. Steam eruptions continued intermittently, with about 400 recorded between 1914 and 1921. These events at Lassen Peak were the last Cascade eruptions for nearly 60 years until Mount Saint Helens blew in 1980.

Today Lassen Peak remains as an active volcano with fumaroles (steam vents), hot springs, and mud pots present throughout Lassen Volcanic National Park. Fortunately, the area around the peak is only sparsely populated, so risks to people and property from any future eruptions are relatively low. The greatest risk is likely to be ash fall, and this would depend on the direction the wind is blowing at the time of an eruption.[3] The USGS has published a map based on detailed mapping delineating the different types of volcanic hazards around Lassen Peak (see figure 4.5).[4]

Cinder Cone

Cinder Cone lies about 10 miles northeast of Lassen Peak and was initially built up to a height of about 700 feet. The most recent geologic studies of the cone indicate that it was formed during two eruptions that occurred in the 1650s and is a complex consisting of several different lava flows as well as several cinder and scoria cones. There was also considerable ash discharged by the original eruption that was deposited as far as 8–10 miles from the vent. While there were older reports of an eruption in the early 1850s, more recent work by USGS scientists confirms that Cinder Cone underwent only a single period of active vulcanism in the period between 1630 and 1670.

Chaos Crags

Chaos Crags formed about a thousand years ago, making them the youngest group of lava domes within Lassen Volcanic National Park. The group of eruptions that formed them make up just one of three instances of Holocene (last ~11,700 years) activity within the greater Lassen Volcanic Center. The lava domes are located about 1.5 miles north of Lassen Peak. In 1974, after a six-year study by the USGS, the most heavily used portion of Lassen Volcanic National Park, the Manzanita Lake area, was closed. The threat of an eruption of hot clastic or fragmented material and of the triggering of large rockfall avalanches from the steep domes in the Chaos Crags area by an eruption or earthquakes led to the decision. Recent mapping indicated that high-velocity, air-cushioned avalanches of rock debris had occurred about 350 years ago when one of the domes collapsed, and the debris had traveled about 4 miles at a speed of about 100 miles per hour, flattening the forest along the way. The cause of the failure is not known for certain, but it may have been a large earthquake. The most intensively used portion of the park is only 1.5 miles from the base of the crags, including Manzanita Lake, which was formed as the rock avalanche dammed Manzanita Creek. The area covered by the debris from that ancient event is known today as Chaos Jumbles.

That 1974 USGS report concluded that "some of the most catastrophic geologic events of the recent past resulted directly or indirectly from volcanism at the site of Chaos Crags."[5] Pyroclastic flows and avalanches from the Crags reach the area where parts of the Manzanita Lake visitor center facilities are located. With any resumed eruptions, including pumice, pyroclastic flows, or rapidly moving rockfalls, rapid evacuation of the area would reduce the major threats to visitors. Future pyroclastic flows could extend at least nine miles downslope into the valley floors. The area affected by any tephra or volcanic ash would depend upon wind velocity and direction. The most dangerous events would be sudden rockfalls or avalanches, such as the one that left Chaos Jumbles 350 years ago, simply because they could occur with virtually no warning, affecting one of the most heavily used areas of the National Park.

Mount Shasta

Mount Shasta is the second-highest peak in the Cascades, at 14,179 feet, and is considered a potentially active volcano, like virtually all of the other peaks in the Cascade Range. The mountain is impressively large, and its estimated volume of 85 cubic miles ranks it number one among all of the stratovolcanoes in the Cascade Range. It has a complex eruption history (see figure 4.3) and consists of four overlapping and dormant volcanic cones. The earliest evidence of Shasta's origins dates to about 593,000 years ago when lava first erupted on what is now the mountain's western flank. Subsequent flows built the peak higher until between about 300,000 and 500,000 years ago when the entire north side of the peak collapsed, creating a massive landslide or debris avalanche. With a volume of about 6.5 cubic miles, this material covered over 170 square miles. For comparison, this is a little larger than the entirety of Lassen Volcanic National Park and 3.5 times larger than the entire city and county of San Francisco. Activity over the past 300,000 years at Mount Shasta has included long quiescent intervals interrupted by short spans of frequent eruptions.

Shastina is the most intact of the individual peaks and was built about 9,500 years ago mainly by lava flows that reached nearly seven miles south and three miles north of what is now Black Butte. Eruptions from Shastina and Black Butte generated many pyroclastic flows that covered a very large area, about 43 square miles, including large parts of what are now the cities of Weed and Mount Shasta, which together are home to nearly 6,000 people.

Over the past 10,000 years, Mount Shasta has erupted on average about every 800 years (see figure 4.3), but the eruption frequency has increased over the past 4,500 years, with events every 600 years on average. There was an eruption reported by an offshore ship in 1786, but this is disputed, and the most recent dated eruption was about 3,200 years ago. It produced block and ash flows on the volcano's north flank. The mountain has been relatively quiet for at least the past 15 years, with only a handful of small-magnitude earthquakes and no recorded ground deformation. Surveys of the geochemistry of the hot springs and gas discharged from a fumarole near the summit are indicative of a deep-seated reservoir of partially molten rock and make it clear that Mount Shasta is still very much alive. The mountain's eruption history indicates that it can discharge volcanic ash and pyroclastic flows as well as lava, and these materials can be found beneath nearby towns. In the past 1,000 years, more than 70 mudflows have inundated stream channels around the mountain.

Mount Shasta has an explosive, eruptive history, and the worst-case future scenario for residents or visitors to the area would be an eruption of a large hot pyroclastic flow, similar to the 1980 Mount Saint Helens event.[6] Because of the presence of ice and snow, lahars are also possible. Due to the dominant wind patterns, ash or tephra from an eruption would probably blow inland, perhaps as far as eastern Nevada. In addition, there is a chance that a future eruption could result in a collapse of the mountain, as happened when Mount Mazama in southern Oregon collapsed 7,700 years ago to form what is now Crater Lake, but this has been given a much lower probability. The USGS has a monitoring program at Mount Shasta that includes seismometers, cameras, a global positioning system (GPS), and temperature recordings, although earthquake activity has been low and ground deformation has been negligible over the last few decades.

Future eruptions like those of the past could endanger the communities of Weed, Mount Shasta, McCloud, and Dunsmuir, all located at or near the base of the peak. Such eruptions will most likely produce deposits of lithic ash, lava flows, and pyroclastic flows. Lava flows and pyroclastic flows may affect low- and flat-lying ground almost anywhere within about 12 miles of the summit of Mount Shasta, and mudflows may cover valley floors and other low areas as much as 10 to 20 miles from the volcano.[7] The USGS has prepared a hazard map of the Mount Shasta area delineating the locations most susceptible to individual types of volcanic processes (figure 4.6). The California Volcano Observatory has rated Mount Shasta as having a very high threat potential (see table 4.1). As with other natural hazards in California, such as very large earthquakes and tsunamis, Shasta and other volcanoes present a major challenge to geologists and government officials: how to prepare

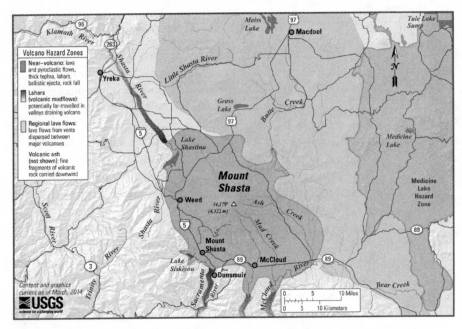

FIGURE 4.6. Simplified volcanic hazards map of Mount Shasta and the surrounding area. *Courtesy of US Geological Survey, public domain.*

for or adapt to a hazard that only occurs very infrequently but, if and when it does, can be catastrophic and deadly.

Medicine Lake Shield Volcano

The volcanic region known as the Medicine Lake Highlands lies about 30 miles northeast of Mount Shasta and extends over an area of about 850 square miles (figure 4.7), an area nearly as large as San Mateo and Santa Cruz counties combined. It lies to the east of the main axis of the Cascade Range and is a gently sloping shield volcano rather than a steep-sided composite cone like the main peaks in the Cascades. A wide shallow caldera is located at the summit of the volcano that contains Medicine Lake.

Medicine Lake Volcano has a history dating back about 500,000 years, and unlike the major Cascade peaks, the eruptions here were generally not explosive and instead were intermittent, effusive flank eruptions of fluid lava. Over the past 13,000 years, flank eruptions produced very extensive lava flows, some covering up to 75 square miles.

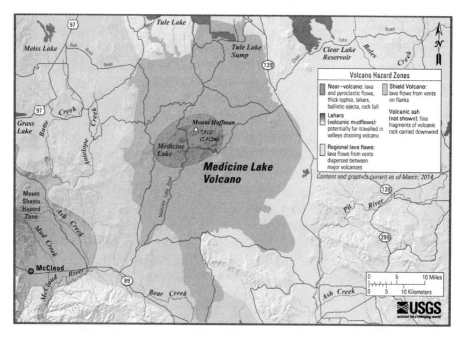

FIGURE 4.7. Volcanic hazard zones around the Medicine Lake Volcano. *Courtesy of US Geological Survey, public domain.*

About two-thirds of Lava Beds National Monument is underlain by one such flow.[8]

Seven of the nine eruptions of the last 5,200 years, however, included an early explosive phase. The two most recent of these sent ash tens of miles downwind during their explosive phases before switching to slow effusion of thick, glassy-looking lava flows (obsidian) forming Little Glass Mountain about 1,000 years ago and Glass Mountain about 950 years ago. Exploratory drilling and geophysical studies reveal a high-temperature geothermal system below Medicine Lake Volcano that is fueled by a deeper zone of magma or partially molten rock. The most significant seismic activity over the last three decades was a series of shallow earthquakes of magnitude 4.1 or less in 1988–89.

Long Valley Caldera/Mono Craters/Inyo Craters/
Mammoth Mountain—Eastern California

About 760,000 years ago, a truly massive volcanic eruption (a "super-eruption") literally blew out roughly 150 cubic miles of molten rock from a depth of about 4 miles beneath the surface in what today is eastern California. That immense volume of material could cover the entire state of California to a depth of nearly 5 feet. Very fast-moving flows of glowing hot ash (pyroclastic flows) covered much of east-central California. Airborne ash was carried nearly 1,000 miles, halfway across the country to Kansas and Nebraska. The hot ash and tephra from that explosive eruption cooled to form the Bishop Tuff, which is known as a welded tuff or ignimbrite. These deposits are massive, up to 600 feet thick just south of the caldera. The ejection of this huge volume of rock and magma led to ground subsidence, creating the Long Valley Caldera, one of the Earth's largest, which is about 20 miles long and 11 miles wide and has as much as 3,000 feet of relief from the caldera rim to its floor.[9] The tectonic reasons or causes for the volcanic activity at Long Valley are mostly unexplained. Domes repeatedly formed in the central part of the caldera following the eruption, occurring at roughly 200,000-year intervals (500,000, 300,000, and 100,000 years ago), followed by a series of younger eruptions. During early resurgent doming, the caldera was filled with a large lake, as much as 1,000 feet deep, that left lakeshore traces (strandlines or bathtub rings) on the caldera walls. The lake eventually overtopped the south rim sometime in the last 200,000 years, eroding the sill and creating the Owens River Gorge.

The caldera remains thermally active, with many hot springs and fumaroles, and has had significant deformation, seismicity, and other unrest in recent years. The geothermal system inside the caldera fuels the Casa Diablo power plant, which generates enough power for about 40,000 homes.

The most recent activity in the area was about 300 years ago in Mono Lake. Both Long Valley Caldera and Mammoth Mountain have experienced episodes of heightened unrest over the last few decades (earthquakes, ground uplift, and/or volcanic gas emissions). As a result, the USGS maintains a dense array of field sensors providing the real-time data needed to track unrest and assess hazards. In May of 1980, a large earthquake swarm that included four magnitude 6 earthquakes—three in the same day—shook the southern part of the Long Valley Caldera, accompanying a 10-inch uplift of the caldera floor. Eight subsequent earthquakes in the same general area in January 1983 indicated continuing activity and the possible intrusion of a dike. Changes in thermal springs and gas emissions were also measured that were all part of the most recent episode of activity in the area. Immediately following the earthquakes, scientists from the USGS began a reexamination of the

Long Valley area and detected other evidence of unrest. Their measurements showed that the center of the caldera had risen almost a foot since the summer of 1979, after decades of stability. This continuing swelling, which by early 2000 totaled nearly 2.5 feet and affected more than 100 square miles, was caused by new magma rising beneath the caldera.

In response to this escalating geologic unrest, the USGS intensified its monitoring program in the Long Valley Caldera and Mono-Inyo Craters volcanic system. An expanded network of seismometers installed in 1982 closely monitors earthquake activity in the area, and other instruments track the continuing swelling in the caldera. Data from these instruments help scientists to assess the volcanic hazards and to recognize the early signs of possible eruptions. In cooperation with the California Office of Emergency Services and civil authorities in eastern California, the USGS has established procedures to promptly alert the public to a possible eruption.

The only modern volcano warning in California history occurred in Long Valley in 1982, when the USGS issued a "Notice of Potential Volcanic Hazard" in response to a series of small earthquakes. The USGS even identified a potential eruption site about two miles from Mammoth Lakes at the edge of Long Valley. This warning made it into the *Los Angeles Times* on May 25, 1982, with a headline of "Mammoth Area Tremors Hint at Volcanism." The Mammoth area had become a popular ski area and extensive development of resort homes and condominiums had taken place. While the earthquakes, which typically precede volcanic eruptions, were very real, nothing followed, but residents were upset and the USGS was generally cast in an unfavorable light. Tourism dropped, dozens of businesses were reportedly closed, and not surprisingly, real estate sales fell because of the media-led volcano scare. Hotels and restaurants in Mammoth Lakes even posted signs reading "Geologists not welcome."

At least one USGS scientist, David Hill, took it upon himself to repair the damage that had been done by the media and opened up communications with residents, including a regularly updated website where any activity or alerts could be posted. This appears to have been the impetus for developing the USGS Long Valley Observatory, which was later to be converted to the California Volcano Observatory discussed earlier.

Initially the objectives of the Long Valley Observatory were to be able to provide residents and government staff with reliable information about the potential hazards posed by any unrest and also timely warnings of a potential eruption. Based on volcano research in many different areas, an increase in seismicity, ground deformation or tilting, and changes in the rate and composition of gas emissions can provide evidence of a potential eruption. Because this part of the eastern Sierras has developed over the years into a large recreation and visitor-serving area, there are far more potential risks to people and property than during any past eruptions.

During the early 1990s, trees began dying off at several places on Mammoth Mountain on the southwest edge of Long Valley Caldera. Studies conducted by USGS and US Forest Service scientists showed that the trees were being killed by large amounts of carbon dioxide gas seeping up through the soil from magma deep beneath Mammoth Mountain. Such emissions of volcanic gas, as well as earthquake swarms and ground swelling, commonly precede volcanic eruptions. When these events precede an eruption of a "central vent" volcano, such as Mount Saint Helens, Washington, they normally last only a few weeks or months. However, such symptoms of volcanic unrest may persist for decades or even centuries at large calderas, such as Long Valley. Recent studies indicate that only about one in six such episodes of unrest at large calderas worldwide actually culminates in an eruption.

Over the past 4,000 years, small to moderate eruptions have occurred somewhere along the Mono-Inyo volcanic chain every few hundred years, and the possibility remains that geologic unrest in the Long Valley area could take only weeks to escalate to an eruption. Nonetheless, geologists think that the chances of an eruption in the area in any given year are quite small. To provide reliable and timely warning prior to an eruption, scientists of the USGS Volcano Hazards Program continue to closely monitor geologic unrest in the Long Valley area.[10]

In April 2006, three ski patrol members died on the slope of Mammoth Mountain in a bizarre accident. Near the peak of the mountain, carbon dioxide and other gases are emitted from a volcanic vent, and they melt the overlying snow. The location has been well known for years, emits the smell of sulfur, and is regularly fenced off by ski patrol members. In this instance, a recent blizzard had dropped over 6 feet of new snow that partially obscured the vent. In the process of moving the fence, there was a collapse of a portion of the snow cover. Three of the skiers fell 21 feet into the cave and died from a lack of oxygen before they could be rescued. One ski patrol member was rescued in time and did recover.

SOME FINAL THOUGHTS ABOUT VOLCANIC HAZARDS IN CALIFORNIA

Unlike most of the other natural hazards discussed in this book that occur frequently and that many residents have likely experienced, erupting volcanoes are far more abstract and remote to most Californians. Active volcanoes are most often perceived as events that occur in places like Hawaii, Iceland, Central America, or Sicily. The most recent major eruption in California was at Lassen Peak, over a century ago, before anyone living in California today was alive. While there are ongoing signs of hot material at some depth beneath the ground surface at all of the volcanic areas discussed in this chapter (small to moderate earthquakes, deformation of the ground surface, hot springs, steam vents, and gas emissions), these are all monitored on a continuing basis by the USGS in order to detect any significant changes that might be indicative of an impending eruption.[11] One of the greatest challenges for both hazard geologists and public officials, however, is how to deal with risk assessment and warnings for events that occur very infrequently but can be extremely hazardous and deadly when they do.

5

Extreme Rainfall and Flooding

While large damaging earthquakes occur somewhere in California perhaps every 20 or 30 years, rivers and creeks tend to overflow their banks with unfortunate regularity, at least in the more mountainous regions of the state. Topography and latitude influence the distribution of precipitation, which generates the runoff that can produce flooding. The state has some historically very wet areas, such as the north coast, where annual rainfall averages over 100 inches in places, and it has some very arid regions, such as southeastern California, where large areas of desert typically get less than 5 inches yearly.

A hundred inches of rain in a year is a lot of water to deal with, although this pales in comparison with what has been recorded in the area generally accepted as the wettest place on the planet—Cherrapunji, India—right at the foot of the highest mountains on Earth, the Himalayas. Topography plays a huge role in rainfall intensity, and monsoon storms in South Asia come off the Indian Ocean laden with moisture, hit the Himalayan foothills, and start pouring out water. This is a place where almost nobody ever leaves home without a heavy-duty umbrella. It rains so much here that they measure it in feet instead of inches, and Cherrapunji holds the world record for one year with 86.8 feet of rainfall! That's almost 3 inches every day, 365 days a year. They are also at the top of the chart for one month of precipitation with 366 inches, or

30.5 feet. That's a foot every day, all month long. Droughts are not a concern in Cherrapunji. The 24-hour record, however, goes to the steep volcanic island Reunion, in the tropical waters of the Indian Ocean, with an astonishing 73 inches, or just over 6 feet in a single day.

At the other precipitation extreme, parts of the Atacama high desert in Chile, sandwiched between the Chilean coastal range and the Andes, have never recorded any rainfall in historic time. They don't sell many umbrellas or rain boots in the Atacama Desert.

At times it seems that California oscillates between droughts and floods, with the former becoming more frequent in recent years as global climate has continued to warm. Yet for many from the Midwest and East, particularly in the post–World War II years, California seemed like paradise. Migration led to an explosion of the state's population, doubling in the 20 years between 1945 and 1965, from 9.3 million to 18.5 million. By 1985, 20 years later, the population had grown another 43 percent to 26.4 million. And everyone needed water, a lot of it—for lawns, swimming pools, golf courses, and most of all for agriculture, for the fruits, vegetables, and nuts not only for those of us who live here, but also for the rest of the nation. Because of the state's climate and fertile soils, California now produces the great bulk of America's fruits, vegetables, and nuts. We are the sole producer of the country's almonds but also grow 99 percent of the walnuts and artichokes, 97 percent of the kiwis and plums, 95 percent of the celery and garlic, 90 percent of the Brussels sprouts, 89 percent of the cauliflower, 71 percent of the spinach, and 69 percent of the carrots, and the list keeps going. It's a long list and they all require water, which is getting harder to come by.

As development took place in the state to accommodate all of these new residents and their needs, it began to encroach more and more onto flood-prone lands. Roughly 20 percent of California communities are vulnerable to flooding due to topography, overflowing rivers and streams, rapid snowmelt, levee failure and dam collapse, or some combination of these. There is a relevant quote here from a quintessential California author, John Steinbeck, in *East of Eden* (1952), which is as appropriate today as it was 70 years ago: "And it never failed that during the dry years the people forgot about the rich years, and during the wet years they lost all memory of the dry years. It was always that way."[1] Most Californians have forgotten about the droughts and also lost track of the floods, even though both have been parts of the Golden State's history for as long as this chunk of land has existed.

THE RECORD FLOODS OF THE WINTER OF 1862

The term *atmospheric river* has now become a common explanation for many of the storms and floods that have regularly drenched and drowned California during historic time. These aerial rivers are massive streams of airborne water vapor that can be 250 to 500 miles wide and hundreds of miles long and that approach California or the West Coast from the Pacific Ocean and can carry massive volumes of water (figure 5.1). Without much doubt, the incessant rains that saturated the western states during the extreme winter of 1861–62 were a product of this newly recognized and named phenomenon. The rain began to fall on Christmas Eve in 1861 and continued relentlessly for the next 43 days. Throughout the entire West, from what was to become the state of Washington to northern Mexico, and from the coast inland to Idaho, Nevada, Utah, Arizona, and New Mexico, the skies just opened up and kept dumping water.

The effect of weeks of continuous rain in the lowlands combined with snow in the higher elevations was amplified by intense warm storms that melted much of the winter snow, contributing further to

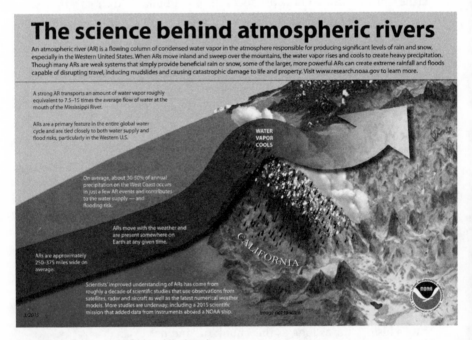

FIGURE 5.1. *Courtesy of NOAA, public domain.*

runoff and flooding. Normally mild streams and rivers that drained from the Sierra Nevada turned into raging torrents that destroyed mining towns and carried away complete communities. Devastating floods occurred throughout California, certainly the worst in history, creating a huge lake—described as being 300 miles long and averaging 20 miles in width and up to 30 feet deep—that covered 5,000 to 6,000 square miles of the vast Central Valley. Fatalities included an estimated 4,000 people and the drowning of a fourth of the state's 800,000 cattle.

The new capitol in Sacramento, which had been built at the confluence of the Sacramento and American Rivers, was completely underwater for 60 days. While a levee had been built along the west side of the city to protect it from flooding, floodwater entered Sacramento from the higher land on the east, with the levee now acting as a dam to keep the water in. The water came in and it stayed. Lower floors of virtually every building in the city, whether residences or businesses, were flooded.

On the morning of January 10, 1862, Leland Stanford, California's governor-elect, was preparing to walk the short distance from his Sacramento mansion to the new capitol building for his inauguration. That wasn't going to happen that morning, however, because the streets of the city were completely underwater. Stanford used a rowboat to get to the capitol instead. For weeks, boats were the only method of getting around in Sacramento.

A passenger on a riverboat heading up the Sacramento River during the flood wrote the following account:

> I was a passenger on the old steamer *Gem*, from Sacramento to Red Bluff. The only way the pilot could tell where the channel of the river was, was by the cottonwood trees on each side of the river. The boat had to stop several times and take men out of the tops of trees and off the roofs of houses. In our trip up the river we met property of every description floating down—dead horses and cattle, sheep, hogs, houses, haystacks, household furniture, and everything imaginable was on its way for the ocean. Arriving at Red Bluff there was water everywhere as far as the eye could reach, and what few bridges there had been in the country were all swept away.[2]

One man told of his journey trying to get to San Francisco from San Jose. Roads either were underwater or had turned to mud. The trip took 36 hours, and as he traveled north, he saw bodies being carried along by the flooded creeks. It took months for the floodwaters to drain out and into San Francisco Bay. The state legislature was even relocated for a while from Sacramento to San Francisco, which with its many hills was more or less immune from runoff and flooding. Following the 1861–62 flood, Sacramento made the decision to raise the entire town. Every home and business was elevated by 10 feet in the subsequent months, some by jacking up existing buildings and some by adding fill beneath them.[3]

For the most part, California residents in that era were a hardy group and regularly rebuilt after each calamity. They seemed to possess a short disaster memory, as most residents do today. The 1861–62 floods are now ancient history, however, and in the years after that epic flood the state's developers, farmers, and industrialists have been busy building out over those same areas that had been submerged many years earlier. Most of the homeowners in these areas today have no idea of the risks that they bought into, nor do the insurance companies that are under-writing them. Our historic record is short, however, so until recently we had no knowledge of how frequent or infrequent an event like the 1861–62 floods was. Was this the 500- or 1,000-year flood? If so, this would appear to significantly reduce the probability of such an event happening again soon. To add further uncertainty, as the climate is changing, the patterns of rainfall, streamflow, and flooding will almost certainly be different than in the past.

The winter of 2022–23 surprised many Californians. After two de-cades of drought, the skies opened up again, dropping record snowfall in the Sierras and a lot of rain over the rest of the state. Reservoirs filled and low-lying communities flooded. Tulare Lake, referred to well over 150 years ago as the largest body of fresh water west of the Mississippi River, partially reappeared. Annual Sierra snowmelt, draining into undammed streams, historically filled as much as 690 square miles of the southern end of the San Joaquin Valley floor, an area nearly as large as all of Con-tra Costa County. But the ancient Tulare Lake has been mostly dry for the past century due to a combination of dams, levees, and canals that controlled and distributed river water. Farmers claimed the dry lakebed, initially for annual crops like tomatoes and cotton, which were gradually replaced with permanent crops such as almonds and pistachios. With this winter's record rainfall and runoff, Tulare Lake partially reformed to fill an area greater than any time in the past 150 years, drowning those permanent orchards and displacing the cattle ranches. There are many lessons to be learned from California's hydrologic history, but short-term gains have often trumped the history of flooding.

Scientists are good detectives, however, and realized not long ago that the major floods of the past carried huge amounts of sediment in addition to immense volumes of water. When the rivers and streams emptied into San Francisco Bay and the surrounding marshes, they left distinct sandy and silty sediment deposits behind. These sediments were slowly covered over by clays that are more typical of these low-energy estuarine environments. Sedimentologists (those Earth scientists who study sediments) have extracted sediment cores from a number of different locations around the bay and have determined through carbon-14 dating that huge floods, comparable to those of 1861–62, occurred in Central California around the years 1100, 1400, and 1650. Similar studies in Southern California, where two-thirds of the state's people live today, looked at sediment cores collected from the offshore Santa Barbara Basin and documented massive floods in (approximately) 212, 440, 603, 1029, 1418, and 1605. Given the uncertainties in precise carbon-14 dating, the last three years match reasonably well with the three very large prehistoric floods in Central and Northern California as recorded in the marsh deposits around San Francisco Bay. Combined with similar studies elsewhere in the state, evidence indicates that these mega-atmospheric rivers and their associated deluges occur every several hundred years, more or less.

We can expect these events to occur in the future, but the big difference will be in the state's exposure to flooding. To begin with, California had just 380,000 residents in 1860, but by 2023 those numbers had reached about 39 million, more than a hundred times as many people. Along with those additional people came agricultural and industrial development and the growth of towns and cities, often on historic floodplains.

Recent findings from investigations of the history of atmospheric rivers, now that this phenomenon is better understood, indicate that there are also significant benefits to California from these periodic downpours. Between 1950 and 2010, it is believed, atmospheric rivers delivered 30–50 percent of California's water, and this usually came in just 10 days each year. In addition, and perhaps this shouldn't be surprising, these storm events also produced over 80 percent of the flooding along the state's rivers. It is not completely clear, however, how climate change may impact the frequency and intensity of atmospheric rivers in the future. What we do know is that as the atmosphere warms, it can hold more moisture, which suggests that atmospheric rivers of the future could carry more water and be potentially more destructive.

Scientists from the US Geological Survey (USGS) have recently run computer models or scenarios for how a megastorm similar to the

1861–62 event—except using only 23 days of continuing rainfall instead of the 43 days of that event—would affect California. Their results should give us some pause for thinking about the future impacts of atmospheric rivers and indicate that sustained flooding could occur over most lowland areas of both Northern and Southern California. In addition to having over a hundred times more people than in 1860, the state now has huge urbanized areas in Southern and Central California and the accompanying vast expanses of impermeable surfaces (streets and freeways, driveways and parking lots, roofs and sidewalks, etc.), which cause rainfall to run off and intensify downslope, increasing downstream flooding. This megastorm flooding could lead to the evacuation of about 1.5 million people. Damage and destruction from a combination of flooding and debris flows in steeper terrain could lead to $400 billion in property and agricultural damages and losses and would likely result in a large number of fatalities. While our improved technology will allow for the prediction of such an event perhaps several days or a week in advance and allow for some reduction in the loss of life, the atmospheric river will still arrive and soak the state for a prolonged period of time. People can move, but homes, businesses, and towns cannot.

1928 SAINT FRANCIS DAM DISASTER, SAN FRANCISQUITO CANYON, SANTA CLARA RIVER

The catastrophic failure of the Saint Francis Dam, located in San Francisquito Canyon in northwestern Los Angeles County, was an important event that focused public attention, at least for a short time, on the need for very careful geologic considerations in the siting of large dams.[4] This dam was built between 1924 and 1926 to serve Los Angeles's growing water needs and was part of a then rapidly expanding network of canals, pipelines, dams, and reservoirs being hastily constructed to bring water to Southern California, an arid region with very little water of its own. The dam and reservoir were located about 40 miles northwest of the city center of Los Angeles and about 10 miles north of the present city of Santa Clarita.

Constructed of solid concrete, about 185 feet high, it stretched for nearly 700 feet across the canyon (figure 5.2). While some modifications were made during construction, the ultimate capacity of the reservoir was about 38,000 acre-feet, or the equivalent of 19,000 Olympic size swimming pools—in short, a very large volume of water. The dam

FIGURE 5.2. The Saint Francis Dam prior to failure. *By H. T. Stearns, courtesy of US Geological Survey, public domain.*

failed nearly instantaneously on the evening of March 12, 1928, less than a year after the reservoir was completely filled and when the water level was within 12 inches of the top of the dam.

There were different but equally unstable rock types on either side of the dam. Both ancient and recent scars showed that landslides had occurred in schist (a metamorphic rock consisting largely of mica, a very planar mineral) on one side. Sedimentary rock on the opposite abutment was dropped into water during investigations after the dam failure and was found to disintegrate rapidly. While apparently stable and resistant in the arid climate at the dam site, this rock was highly unstable when saturated. A prominent and legendary geology professor at the University of California, Berkeley, Andrew Lawson, reportedly brought a piece of the same rock into a geology class immediately following the dam failure and dropped it into a beaker of water at the front of the classroom. As the students watched the rock disintegrate to a mush, Lawson bellowed, "Pea soup! They built the damn thing on pea soup!"[5]

Adding to the problems of construction on two unstable rocks types, a fault contact between the two different formations was marked by a wide zone of sheared and broken rock and passed directly under the dam. This fault also appeared on the existing fault map of California. On the day of dam failure, a geology graduate student from the California Institute of Technology (Caltech) had been following the trace of the fault as he hiked up San Francisquito Canyon. Late in the afternoon he rounded a bend in the canyon and looked up to see the nearly 200-foot-high Saint Francis Dam blocking his path and sitting directly over the fault. His first thought was that he had made some mistake in his mapping of the fault, as surely nobody would build a major dam across a fault. He had planned to camp in the valley that night below the dam, but when it started raining, he made a wise decision to return back home to Pasadena instead.

That evening, three minutes before midnight, the dam failed catastrophically, emptying the entire 38,000 acre-feet of stored water (12.4 billion gallons), which surged down San Francisquito Canyon as a flood wave, initially 140 feet high (figure 5.3). Within five minutes of the collapse, the flood was still 120 feet high and had traveled a mile and a half downstream at an average speed of 18 miles per hour, taking the lives of 64 of the 66 workmen and their families working on a powerhouse. The flood very quickly inundated parts of present-day Valencia and

FIGURE 5.3. The remains of the Saint Francis Dam following the March 1928 failure, looking upstream at the reservoir floor. *Courtesy of US Geological Survey, public domain.*

Newhall and then, farther downstream, Castaic Junction, Fillmore, Barsdale, and Santa Paula before discharging the water, victims, and debris into the Pacific Ocean, 54 miles downstream from the reservoir.

The exact number of victims remains unknown nearly a century later. While the official death toll in August of 1928 was 385, the remains of victims continued to be discovered every few years until the mid-1950s. Many victims were carried out into the ocean and were never recovered. Others were washed ashore, some as far south as the border with Mexico, 175 miles away from the river mouth. The death total was later estimated to be closer to 500. This remains the third-highest number of fatalities from a geologic disaster in California history, following the 1861–62 flood and the 1906 San Francisco earthquake and fire.

A careful study of the site following dam collapse, including the positions of several large fragments of the dam, indicated that failure occurred initially along the eastern abutment because of a massive landslide within the schist. The entire eastern side of the dam failed with the slide. The center of the dam was also knocked loose by the landslide movement. With the strength of the arch design now gone, the force on the west side led to failure of the wing wall. Complete failure was catastrophic, and the entire reservoir emptied in less than 70 minutes. While some parts of the collapsed dam remained where they had fallen, the largest piece, weighing about 10,000 tons, was found three-fourths of a mile downstream.

A number of inquiries and studies followed the disaster, focused on dam failure, and many reports were subsequently published. The consensus of geologists and engineers was that the dam failed because of adverse geological conditions at the site that were either unrecognized or ignored.[6]

MARCH 1938 LOS ANGELES FLOOD—LOS ANGELES, ORANGE, AND RIVERSIDE COUNTIES

In early March of 1938, two back-to-back storm systems hit Southern California with record rainfall for several days. Today these have been recognized and better understood as powerful atmospheric rivers coming off the Pacific Ocean to the west. For the normally arid Southern California region (the long-term annual average rainfall for Los Angeles is 14.2 inches), the total rainfall for these two storms was way off the chart. Downtown Los Angeles received 5.88 inches on the wettest single day, March 2. Most of Southern California was deluged with almost a year of precipitation in just a few days. Total damages reached $1.6 billion (in 2022 dollars), and the flooding and debris flows took the lives of 115 residents, destroyed 5,600 homes and businesses, and damaged 1,500 more.

The greater Los Angeles basin is backed by the San Gabriel Mountains, which form a topographic obstacle for moisture-laden storms coming in off the Pacific Ocean. As the clouds rise approaching the mountains, the air begins to cool, moisture condenses to water, and the rain starts to fall (see figure 5.1).

A number of factors, some natural and some human, contributed to the 1938 floods as well as those that came before and that have come after. The San Gabriel mountains are steep and, due to both the hard, dry soils and the relatively impermeable underlying bedrock, are prone to rapid runoff. As people migrated to Southern California, the population continued to expand, with all of the associated effects of urbanization on runoff and flooding. In 1900, Los Angeles County had just 170,250 residents, which constituted 11.5 percent of California's total population. By the time of the 1938 deluge, the population had increased 8.5 times, or had grown to about 1,444,000, and amounted to 40 percent of the entire state's residents. The alluvial fans at the base of the mountains and the flat-floored valleys, which had been formed by centuries of flooding along the meandering streams, were gradually subdivided and covered with homes and schools, businesses and industries, streets and freeways, and sidewalks and parking lots.

The hardest hit region in the March 1938 storms was the San Fernando Valley, where many communities were developed during the economic boom of the 1920s in flat, low-lying areas that had formerly been used for agriculture. Many homes had even been built in old streambeds that hadn't been flooded for years, giving a false sense of security to developers and homeowners. Formerly exposed ground that historically absorbed the rainfall was now impervious, increasing the volume and velocity of discharge. The runoff from the surrounding mountains led to the overflow of the normally dry Los Angeles, San Gabriel, and Santa Ana Rivers, which flooded much of the coastal plain of the San Fernando and San Gabriel Valleys. Stream channels were eroded, leading to multiple bridge failures from the force of the floodwaters or from the impacts of floating buildings and other debris (figure 5.4). The collapse of a bridge over the Los Angeles River at Compton led to 10 deaths. In the San Fernando Valley, five people died when the Lankershim Boulevard bridge collapsed at Universal City. Runoff in the mountains also liquefied the soils, generating high-velocity mud and debris flows that

FIGURE 5.4. March 2, 1938, a raging Los Angeles River destroys the Southern Pacific Railroad bridge. *Copyright © 1963 Los Angeles Times. Used with permission.*

destroyed homes and inundated streets in the surrounding mountainous areas. Communication and transportation systems were cut off for days as railroads and streets were buried and power, gas, and phone lines were ripped apart. Some neighborhoods were buried with up to six feet of mud and debris, requiring a major cleanup effort.

The response of the US Army Corps of Engineers and local agencies was ultimately to constrain the great majority of the region's streams in concrete flood control channels and also to build a number of dams and debris basins in the smaller drainages to control both floodwaters and mud and debris from the steep watersheds. While the 1938 flood is referred to statistically as a 50-year flood, subsequent floods (1969 and 2005) were actually larger in discharge volume, but significant flooding was greatly reduced by the flood control projects constructed in the aftermath of the 1938 disaster.

THE TROPICAL CYCLONE OF SEPTEMBER 25, 1939

Southern California isn't known for tropical cyclones; these belong along the Gulf Coast and the Atlantic Seaboard, although the west coasts of Mexico and Central America do frequently get hit. In late September of 1939, however, as Los Angeles residents suffered through an unusual heat wave, four tropical cyclones were developing in the eastern North Pacific. The largest of these, which was unnamed, made landfall at San Pedro on September 25 and dropped 5.7 inches of rainfall on Los Angeles, 40 percent of the city's average annual precipitation in just a few days. This was the only tropical storm to hit California during the entire 20th century. Winds of up to 65 mph battered the area, damaging structures, boats, and utility lines, with losses exceeding $42 million (in 2022 dollars). Damage to crops in the inland farms reached an additional $9 million.

Flooding from this deluge led to 45 fatalities. An additional 48 lives were lost at sea as vessels were caught completely off guard. The coast was hit hard by large waves, with shoreline homes flooded from Malibu to Huntington Beach. Six people were drowned on beaches alone. Vessels were sunk, capsized, or blown onshore, where they smashed against piers, breakwaters, and the coastline. Half of the pier in San Clemente was destroyed at a loss of about $21 million (in 2022 dollars). Twenty-four lives were lost on a vessel called the *Spray* as it attempted to dock at Point Mugu, where the pier was destroyed. There were just two survivors, a man and a woman, who were able to swim to the beach and then proceeded to walk five miles to the city of Oxnard. Fifteen more drowned on board a fishing boat named *Lur*.

The Los Angeles River, normally a trickle at that time of year, became a raging torrent and flooded many low-lying areas, stranding thousands of people in their homes. Water depths reached two to three feet in Los Angeles and Inglewood, flooding buildings and stalling cars. Along the shoreline, windows in Long Beach were shattered by the high winds. At Belmont Shores, storm waves undermined 10 homes, which were then washed into the ocean.

DECEMBER 1955 FLOODS

The floods of December 1955–January 1956 in the far western states were at the time, in the words of the USGS, "the greatest in the area in the history of streamflow records," in large part because of the unusually large area involved—the western half of Oregon, the western third of Nevada, one-third of western Idaho, minor parts of Washington, and the northern two-thirds of California.[7] Of overall significance were the record-breaking peak river discharges and the extensive areas flooded.

Total damage exceeded $2 billion (in 2022 dollars). The actual calculated losses, however, were only part of the larger picture of devastation. There was incalculable indirect economic impact and disruption from reduced crop production, losses in business volume, salaries and wages, traffic delays and rerouting, and reduced tourist trade, to name a few. The western states were actually hammered by a series of storms, three that occurred from December 15 to 27 and three more from January 2 to 27. The coastal area of Northern California experienced measurable rainfall on 39 of the 44 days between December 15 and January 28, with several stations recording over 60 inches of precipitation during this 44-day period.

The Christmas flooding of December 1955 was one of the greatest disasters of its kind up until that time in California. It's necessary to go back nearly a century in the state's history to find a record of any comparable deluge. That was the great winter flood of 1861–62, about which much has been written, but not a lot of factual information was recorded at that time.

The 1955 storm, which at that time was labeled a Pineapple Express but today would likely be ascribed to an atmospheric river, was actually a series of storms. Warm temperatures raised the snow level to 9,000 feet as the persistent rain saturated the ground and melted much of the region's promising 3-foot snowpack. The rains affected virtually the entirety of Northern California, from the central Sierra Nevada to the coast and as far south as the Santa Cruz Mountains. A statewide disaster was declared, with the storms resulting in 74 deaths and approximately $2.1 billion in economic losses (in 2022 dollars).

On December 20, Humboldt County was hammered with 15.3 inches in just 24 hours. The Eel River on the north coast saw the greatest discharge of record up to that time. Its previous high flow of record for 45 years, at the most downstream gauging station (Scotia), was 345,000 cubic feet per second (cfs). On December 22, 1955, maximum discharge reached 541,000 cfs, 57 percent higher than the maximum previously recorded.[8] That volume of water could fill six Olympic size swimming pools every second.

The Central Valley also experienced very high rainfall and record streamflows during these storms. The heaviest 24-hour rainfall was recorded on December 20, when 15.3 inches fell in Shasta County. At Red Bluff, the Sacramento River peaked at 291,000 cfs, more than doubling the previous high discharge of 116,000 cfs. The storm's toll on Sutter County was among the greatest in the state.

The small Central Valley town of Yuba City lies at the confluence of the Yuba and Feather Rivers, which were protected by what were believed to be strategically placed levees. Fortunately, the levees of the nearby town of Marysville held during the storm, although some of the floodwater still made its way into the town. The levee south of Yuba City wasn't as fortunate, however. At 12:04 a.m. on Christmas Eve 1955, a 21-foot-high wall of water poured in through the town and washed the 5th Street bridge downstream. With the collapse of the levee, downtown Yuba City quickly found itself sitting under 8 feet of water (figure 5.5), with its 10,000 residents immediately fleeing to find dry land. In the midnight darkness, the town would go into complete chaos, with families scattering from their homes to escape the deadly floodwaters. Ninety percent of Yuba City was underwater, as were the farmlands in the southern Yuba City basin.

In the end, 40,000 people would evacuate the surrounding area, with over 600 needing to be rescued by boat or helicopter from the tops of houses and tree branches by Army units dispatched from Camp Beale. Thirty-eight people died in the floodwaters. Following the flooding, the

FIGURE 5.5. Flooding of downtown Yuba City on Christmas Day, December 1955. *Courtesy of Sutter County Library.*

US Army Corps of Engineers stated that this area is "the most prone to intense flooding of any river valley in the United States." If it weren't for the efforts of the military, many more may have died.

Eventually, the town would dry out and rebuild. Today, more than 66,000 people live in the small town, with memories of flooding long in the past and new, improved levees hopefully providing future flood protection. But for the older generation of Yuba City residents, the memory of their town underwater is something that will never be forgotten. As the climate is changing, however, the historic precipitation and stream-flow records, which have been used to design flood control structures, may no longer be reliable indicators of what may be coming in the future. This is a very serious concern that needs to be carefully considered and kept in mind when new flood control structures are being planned or older ones are being repaired or rebuilt.

Guerneville, on the Russian River, has been flooded more times than most residents can probably remember. In 1955, the downtown area was completely underwater, with National Guard amphibious vehicles coming to rescue many of the threatened residents. There are no simple solutions for this small town on the river's floodplain, and because of its elevation and location, floods will continue.

On the west side of San Francisco Bay, between the cities of Palo Alto and Menlo Park, San Francisquito Creek had been flooding for centuries, but it didn't really seem like a problem to most residents. Three days before Christmas in 1955, however, it became a major problem for everybody living in the new housing developments that surrounded Greer Park. During earlier floods, the area had been mostly undeveloped apricot orchards, but by the mid-1950s new home construction was booming. Because the winter had actually been a dry one in Northern California, many people were happy to finally see rain fall on December 9, bringing some relief to the parched crops of the Santa Clara Valley. But continuing rain was reported for December 18, 19, 20, and 21, with more on the way. On December 22 it rained all day and into the evening as San Francisquito Creek continued to rise.

Wood and debris carried by the creek was caught under the bridge over Bayshore Highway, forming a dam that backed up the creek. Rising waters jumped the levees on the Palo Alto side of the creek just before midnight, opening a 20-foot gap in one levee wall. Creek water rushing downhill was joined by runoff from impermeable streets and gutters along Greer Road and flowed into the surrounding homes along the way. At the dead end of Seale Canal near Greer Park, water spread out and inundated homes throughout the low-lying subdivisions south of the embarcadero. By 1:45 a.m., evacuees were everywhere. Flooded streets kept people from leaving in their automobiles, and eventually more than a thousand people were forced to abandon their homes as emergency trucks and buses mobilized to transport flood victims. At 1:50 a.m., a local shop teacher managed to open up Jordan Junior High School, which became a shelter for the rest of the night. At one point more than 600 evacuees crammed into the school, most of them without shoes, socks, or dry clothing.

In the end, no one was seriously injured here in the 1955 flood, but damage was extensive. More than 650 Palo Alto homes were flooded, totaling $12.3 million in damage (in 2022 dollars). Many residents spent Christmas holidays cleaning mud from living rooms, salvaging damaged furniture, and shoveling out driveways and garages.

On the opposite side of the Santa Cruz Mountains, the San Lorenzo River topped its banks as it entered the city of Santa Cruz, which had been built on the river's floodplain. The river's watershed is characterized by steep slopes, and the total length of the San Lorenzo from the headwaters to the mouth is only about 29 miles, so runoff from rain falling on the highest parts of the drainage basin reaches downtown Santa Cruz within about 12 hours. Boulder Creek in the upper watershed was inundated with 18.3 inches of rain in the three days before Christmas. The river crested in Felton at 22.55 feet, with a peak flow of 30,400 cfs. (This flow would fill an Olympic size swimming pool in less than 3 seconds; put another way, if you built a 10-foot-high waterproof wall around a football or soccer field, this river flow would fill it up in 14 seconds.)

Overflow occurred from the headwaters to the mouth, and water levels topped all previous high-water marks along Kings, Boulder, Two Bar, Bear, and Zayante Creeks in the upper San Lorenzo basin. The persistent heavy rains and floodwaters loosened and scoured out riverbank trees and logs, floating them downstream, where they again became lodged at channel constrictions such as low bridges. The many logjams diverted the high-velocity river flows, undercutting and scouring bridges and road fills as well as homes and other structures. Nearly 300 acres of land in the watershed from Felton upstream were flooded. The total estimated San Lorenzo River basin damage was $98 million (in 2022 dollars), of which $84 million was within the city limits of Santa Cruz.

The following reports from the local paper (the *Santa Cruz Sentinel*, Christmas 1955) provide the most graphic and somewhat literary descriptions of the flooding and the resultant damage.

The torrenting San Lorenzo River which spread death and destruction through Santa Cruz dealt severe damage along its path in the San Lorenzo Valley. Nearly 100 homes along the usually placid river were destroyed— some were swept down by the roaring flood. Along the tributaries of the San Lorenzo more homes were torn and splintered. Many families are still isolated with dwindling food supplies. In the dark, murky night of Thursday, the waters rose over the bank and charged with unbelievable force through the thick forest resort parks. It crammed five-room houses against trees, twisted others off their foundations.

Felton Grove . . . was nearly wiped out. Twelve persons were taken out of the flooded area in boats. Five cabins were carried away, some 20 were destroyed, crushed by the rushing current. The sight of Gold Gulch Park, a subdivision of private homes in Tanglewood [just downstream from Felton] was startling. Looking down toward the river from the highway, I could see a tumbled muddy mess of expensive mountain homes, shoved here and there like a puzzle maze. Thirty homes were pushed off their foundations. Three were reported gone down the river. . . . One house was rammed by a large tree trunk, which had rushed down the river and lodged inside the house.

Santa Cruz braced itself for another night of wet terror after undergoing the greatest flood in the recorded history of the area last night. The still-rampaging San Lorenzo River roared from its banks at 9:30 last night in a destructive surge that drowned the city's central districts in up to 10 feet of water.

Pacific Avenue and the residential and industrial areas bordering the river were inundated by smashing torrents of water that tore up stores, shattered homes and forced hundreds upon hundreds of residents to flee to higher ground. A number of persons were reported swept from sight by raging currents that smashed toward the sea on both sides of the river [total death toll in the city was actually only five persons].[9]

Seventy-four lives were lost in California's 1955 floods. While many communities and cities were heavily damaged, perhaps the site of greatest destruction and loss was in Yuba City and Marysville, where 38 people were killed and nearly 300 homes were destroyed and more than 1,500 damaged. The destruction and losses were widespread throughout Northern California, and the above accounts only represent a few examples of the impacts on many individual communities. This was the largest flood in the state since the infamous 1861–62 event.

BALDWIN HILLS DAM FAILURE—1963

Not all of the state's flooding has been solely from streams overtopping their banks from prolonged or intense rainfall. There have been some colossal floods from the failure of dams and levees, the most disastrous being the 1928 Saint Francis Dam failure described earlier. The Baldwin Hills Reservoir, southwest of downtown Los Angeles, is another example. This dam was constructed between 1947 and 1951 on the northeast flank of the Inglewood oil field. The reservoir was cut out of the top of a hill consisting of poorly consolidated silts and sands, which were extremely loose and erodible. The active Newport-Inglewood fault zone passed within about 500 feet of the reservoir site, and preconstruction investigations revealed fault traces beneath the site. There were also measurements of nearly 10 feet of subsidence of the ground surface about a half mile west of the reservoir, likely from some combination of petroleum withdrawal and tectonic activity along the fault. Dam and reservoir construction took place in a post–World War II climate of rapid population growth in the greater Los Angeles area and an increasing demand for water storage facilities, which in at least two cases led to quickly planned and constructed projects.

The Baldwin Hills dam and reservoir were built by the Los Angeles Department of Water and Power even though the known engineering geology before and during construction indicated that the site was in an area of known active subsidence close to a major fault, and that the site itself was crossed by parallel active faults. The reservoir lining and dam incorporated some of the most modern design innovations of the day, including a reservoir lining and an elaborate drainage system intended to deal with the potential foundation issues. On December 14, 1963, with less than four hours of warning, the dam failed (figure 5.6). The resulting flood claimed five lives and destroyed 277 homes, with damages totaling $147 million (in 2022 dollars) and an immediate water shortage for half a million people.

Immediately following failure, a continuous crack with a vertical displacement of two inches was visible crossing the floor of the reservoir, which ruptured the supposedly impermeable lining and was subsequently found to match with one of the faults. Years of investigations and lawsuits following the disaster ultimately concluded that fluid injection used to repressurize the nearby oil field to recover more oil led to activation of the faulting and subsidence.[10]

CHRISTMAS FLOODS OF 1964

Nine years after the Christmas floods of 1955, the Pacific coast was hit again with sustained rainfall over Christmas in 1964 from an atmospheric river that produced the highest water levels and worst floods in

FIGURE 5.6. The 1963 Baldwin Hills dam failure following reservoir emptying. *Copyright © 1963 Los Angeles Times. Used with permission.*

recorded history on nearly every stream and river in coastal Northern California. An area of about 200,000 square miles, roughly the size of France, experienced flooding, with overall damage in the western states reaching about $5 billion (in 2022 dollars). The streamflows and areas flooded exceeded those of the 1955 floods. Nineteen people reportedly died, and at least 10 towns were either completely destroyed or seriously damaged. Floodwaters devastated all or parts of over 20 major highways and bridges and inundated thousands of acres of agricultural land, with the loss of about 4,000 head of livestock.

Intense downpours of unprecedented intensity began on December 21, 1964, in Northern California and continued into the beginning of January. Ettersburg on the Mattole River recorded a staggering 50 inches of rain in this period, including 15 inches on December 22. The Standish-Hickey State Recreational Area along the Redwood Highway recorded 22 inches of rain in a single 24-hour period. Thirty-four counties were declared disaster areas, with Del Norte, Humboldt, Mendocino, Siskiyou, Trinity, and Sonoma Counties suffering the greatest damages, more than the other 28 counties combined. The major Northern California rivers—the Eel, Smith, Trinity, Salmon, and Mad—all surpassed flood stage and peaked on December 21 and 22, breaking earlier records.[11] Daily discharge measurements collected near the mouth of the Eel River showed flows increased from 10,000 cfs on December 19 to nearly 700,000 cfs the following week. Translated into more understandable measurements, if you put a 16-foot-high wall around a standard football field to make a very large pool, this Eel River flow would have filled that pool in one second—in short, a massive amount of water.

Sixteen state highway bridges were destroyed by the raging rivers, most of them along Highway 101. Thirty-seven miles of railroad track were also washed away. The town of Klamath was covered by 15 feet of water. Riverside communities like Orleans, Myers Flat, Weott, South Fork, Shively, Pepperwood, and Stafford were all decimated by the 1964 floods (figure 5.7). A few of these were never rebuilt. Larger towns like Rio Dell, Scotia, and Ferndale were heavily damaged. The Eel River watershed seemed to have received the brunt of the storm, with over 22 inches of rain falling in just two days. At the mill town of Scotia along the river, the Pacific Lumber Company lost 40 million board feet of lumber, which along with thousands of old-growth redwood logs, were carried downstream. The logs battered everything in their path,

FIGURE 5.7. Flooding of the town of Weott on the Eel River, 1964. *Courtesy of Greg Rumney, Old Photo Guy.*

knocking homes off their foundations and destroying bridges on their way to the ocean. Over a dozen bridges in Humboldt County were taken out. Damages and losses along the Eel River Valley alone reached almost $740 million. Farther south, the town of Klamath was essentially wiped off the map and under 15 feet of water.

The Central Valley wasn't spared, with about 375,000 acres being inundated. Yuba City's residents were again evacuated, as they were in 1955. The Feather River, Yuba River, and American River and Butte and Cottonwood Creeks all reached record flood stages. Statewide there were 47 fatalities from the 1964 floods. Total damages and losses in California and Oregon reached nearly $5 billion (in 2022 dollars).

THE 1969 WINTER STORMS AND FLOODS

Two days of heavy rain in January 13–14, 1969, were the beginning of a downpour that was to continue for nine days throughout Southern and Central California and was to be followed by several subsequent storms that extended heavy rainfall intermittently through February 25. The areas most affected by flooding included the watersheds of streams draining the central and south-coastal ranges, the southern part of the San Joaquin Valley, and the southern Sierra Nevada foothills from Mariposa Creek north of Fresno to the Kern River. Many of the stream discharge records were the greatest in 30 years and in Southern California were nearly as large as the March 1938 floods. The flow from several streams, the Santa Clara, Santa Ynez, and Salinas Rivers, exceeded the highest levels ever recorded and may have been close to those of the infamous floods of 1861–62. Rainfall measurements indicated that the mountains surrounding Los Angeles recorded 9 to 48 inches of precipitation during an eight-day period. With the ground already saturated, additional rainfall from January 24 to 27 immediately became runoff from the San Gabriel Mountains, producing extreme floods with catastrophic effects. Normally dry Cucamonga Creek became the raging Cucamonga River.

It seemed to be the mountain and foothill communities that were hit the hardest, with some residents seeing water three feet deep flowing through their living rooms, not a normal phenomenon for the normally arid Southern California region. The runoff washed out bridges, along with roads and streets. Rail transportation was interrupted, homes were destroyed, and severe landslides occurred in the mountain areas.

Total precipitation for the two-month period of January and February 1969 reached a maximum of 84.6 inches (just over 7 feet of water) at Mount Baldy, 81.9 inches at Lake Arrowhead, and 78.8 inches at Opids Camp at an elevation of 3,600 feet in the San Gabriel Mountains. At normally dry Los Angeles, the combined two-month rainfall exceeded 21 inches, the largest total since 1884.

Ninety-two lives were lost from the storms and flooding, with about 10,000 people driven from their homes. Total losses were estimated at over $3 billion (in 2022 dollars). Despite the record-breaking or near-record rainfall and floods, damage was minimal in older developed areas that had been protected against inundation and debris damage by well-planned flood control facilities, such as debris basins and flood conveyance channels. These structures prevented an estimated additional $3 billion in damage from occurring. Debris basins in Los Angeles County trapped about 2 million cubic yards of debris, or about 200,000 dump truck loads. Of the 61 debris basins, only 7 were completely filled, and only 3 had debris pass over their spillways into downstream drainages. In general, the flood control channels contained the flood flows within their banks and levees. By contrast, extensive damage occurred in more recently developed areas where flood and debris-control structures had not kept pace with expanded urbanization.

The 1969 storms and flooding were not completely constrained to Southern California, as flooding also took place on the Sacramento River in the delta region, due to levee failure (figure 5.8). The continuing storms that today would no doubt be recognized as atmospheric rivers weren't completely confined to the southern part of the state—35 of California's 58 counties were declared disaster areas, and one news source reported that this was "the worst weather-related disaster in 20th century California."

1982 SAN FRANCISCO BAY AREA FLOODS

The central coast of California, including most of the San Francisco Bay region, experienced prolonged and intense precipitation during January 3–4, 1982, resulting in heavy flood damage. This storm followed two months of abnormally high rainfall and produced 24-hour precipitation in excess of the 100-year event at many stations in the area, equivalent to half of the average annual rainfall within a period of about 32 hours. Some localities received an inch per hour for more than 8 hours, which produced flooding and triggered over 18,000 debris flows throughout 10 bay area counties. These flows swept down hillsides and drainages with virtually no warning, damaged or destroyed at least 100 homes,

FIGURE 5.8. Flooding of a mobile home park on Sherman Island in the Sacramento delta area in 1969. *Courtesy of California Department of Water Resources, public domain.*

and led to 14 fatalities, but these will be treated in more detail in chapter 8 on Landslides, Rockfalls, and Debris Flows.

Thousands of residents left their homes in hazardous areas, entire neighborhoods were cut off as roads were blocked, water systems were destroyed, and power and telephone lines downed. In total, the three-day storm damaged 6,300 homes, 1,500 businesses, and many miles of roads, bridges, and communication lines. This storm system would likely be termed an atmospheric river today; it poured as much as 38.2 inches of precipitation onto the greater San Francisco Bay area over the course of the storms.

Heavy rainfall lasted about 28 hours over most of the Santa Cruz Mountains, with some locations receiving an inch per hour for more than 8 hours. At higher elevations most rainfall measurements exceeded the projected 100-year 24-hour storm values. Compounding this high-intensity precipitation, the pre-January-storm rainfall in November and December had been between 33 and 39 inches, which created conditions favorable for rapid runoff when the January 3–5 storm hit these steep slopes.[12] Many of the county's communities were built, at least in part, on the floodplains of the streams. Downtown Santa Cruz, and Soquel, and sections of Felton, Ben Lomond, Aptos, and Capitola all were built on floodplains.

Logging, road clearing, and home development in the Santa Cruz Mountains contributed to the flooding as logs and other debris piled up at bridge constrictions or blocked culverts, leading to stream impoundment and overbank flooding. Logs and trees up to 70 feet long were carried down the San Lorenzo River and were lodged against narrow bridge openings within the city of Santa Cruz. Flooding here was only averted by cranes that worked throughout the flood peak to remove logs and prevent logjams and flooding. Scouring of the river channel led to the undercutting of the support for one major bridge, which collapsed, leading to the loss of all phone service for the city (figure 5.9), as this was a time before we had cell phones. Farther upstream in the community of Felton, a group of 50 homes and cabins adjacent to the San Lorenzo River, but only 12 to 16 feet above the river bottom, were inundated with floodwater 3 to 6 feet deep, flooding virtually all of the homes and covering automobiles with sediment (figure 5.10). This collection of cabins and homes, known as Felton Grove, is an unfortunate example of the hazards of floodplain development. The houses closest to the river had been flooded 4 times in the previous 12 years and 14 times in the past 46 years.[13]

Soquel Creek has a small, mountainous watershed of 40 square miles and is only 11 miles in length. As a result, streamflows originating in the

FIGURE 5.9. Collapse of the Soquel Avenue bridge on the San Lorenzo River in Santa Cruz due to severe scour of the streambed during the January 3–5, 1982, floods. © 1982 Gary Griggs.

FIGURE 5.10. Sediment covered cars and filled homes in Felton Grove in January 1982. © *1982 Gary Griggs.*

higher parts of the drainage basin reach the community of Soquel very quickly, in fact with almost no warning. A large logjam formed at one of the main bridges in the middle of town, leading to overflow, with the bulk of the flow of Soquel Creek being rerouted through downtown, where water levels reached a depth of five feet. Floodwaters swept through two mobile home parks. Many of the elderly residents in the parks weren't aware of the flood until water began to seep under their doors. The trailers were swept off their piers, and many were damaged by logs and other debris. Only 17 of the 39 mobile homes damaged were determined to be salvageable. In addition, 58 businesses, 21 homes, the firehouse, library, post office, and Grange Hall were all damaged by floodwaters and the mud and debris left behind. The cause of the flood, namely a logjam at a bridge, and the areas flooded were nearly a complete repeat of a 1955 flood.

Farther to the east, Aptos Creek drains a basin of just 16 square miles and extends only 6.6 miles inland from the coast. The watershed is now almost completely forested, after heavy logging was terminated 90 years earlier. Nonetheless, the steep and short drainage delivered what was the equivalent of the 40-year flood. The only significant damage occurred in the lowermost reaches of the creek where homes had been damaged in a 1955 flood. Fill had been imported, the creek channel had been constrained, and homes had been rebuilt on the same floodplain locations. High flows undermined the fill beneath the perimeter foundations, and at least seven homes sustained major damage. Two houses broke in two, due to undermining, with the detached half of one floating downstream. Part of another was lodged against a low downstream bridge. In the months following the 1982 flood, the floodplain was again built up, the creek's course was straightened, and rebuilding again took place in the same locations.

1986 NORTHERN AND CENTRAL CALIFORNIA—SACRAMENTO VALLEY

Three separate storm systems hit California in early 1986. On February 11, 1986, a vigorous low-pressure system drifted east out of the Pacific, creating what was then known as a Pineapple Express, though today it would probably be designated as an atmospheric river. It lasted through February 24, dropping unprecedented amounts of rain on Northern California and western Nevada. While the storms moved southward through the state, Northern California was hit particularly hard, which led to the flooding of several river basins. The 13-day storm over California dropped half of the average annual rainfall for the year. Bucks Lake in the Feather River basin saw nearly 50 inches of rain, and the Russian and Napa River basins were deluged with 40 inches.

Runoff produced record flooding in three streams that drain to the southern part of the San Francisco Bay area. Extensive flooding occurred in the Napa and Russian Rivers. In Napa they recorded their worst 20th-century flood ever. The 20 inches of rain that fell in a 48-hour period led to the evacuation of 7,000 people and the loss of three lives. Total damage, which is always difficult to determine, was estimated at $274 million (in 2022 dollars), with 250 homes destroyed and another 2,500 damaged.

Nearby Calistoga recorded 29 inches of rain in 10 days, creating what was deemed a once-in-1,000-years rainfall event. Records for 24-hour rain events were reported in the Central Valley and in the Sierra Nevada, where 1,000-year rainfalls were recorded. There are, however, very large uncertainties in projecting or determining a 1,000-year rainfall event when the historic record usually only extends back a century at best. In addition, as the climate changes, the historic data may no longer be representative of what can be expected in the years and decades ahead.

The heaviest 24-hour rainfall ever recorded in the Central Valley, at 17.60 inches, occurred on February 17 in the Feather River basin at

Four Trees. Nearly 10 inches of rain fell in an 11-day period in Sacramento. Flood control failures in the Sacramento River basin included disastrous levee breaks in the Olivehurst and Linda areas on the Feather River. Linda, about 40 miles north of Sacramento, was devastated after a levee broke on the Yuba River's south fork, forcing thousands of residents to evacuate (figure 5.11). In the San Joaquin River basin and the delta, levee breaks along the Mokelumne River caused flooding in the community of Thornton and the inundation of four delta islands.

The 13 days of rainfall led to 13 fatalities, over a billion dollars in property damage, and the evacuation of 50,000 people from their Central Valley homes while levees and dams failed under the increased runoff. Mudslides and rockfalls caused by the floods led to several highways being blocked off completely.

JANUARY AND MARCH 1995—NORTHERN AND CENTRAL CALIFORNIA

El Niño conditions during the 1994–95 winter brought heavy rainfall to Northern and Central California, producing floods that led to federal disaster declarations in 57 of the state's 58 counties. Five to 10 inches

FIGURE 5.11. Flooding in Linda from a levee break in the Sacramento Valley in April 1986. *By Michael J. Nevins, courtesy of US Army Corps of Engineers, Sacramento Division, public domain, via Flickr.*

of precipitation per month fell for a three-month period, and over 100 rainfall stations recorded the largest amount of one-day precipitation in the stations' history. Along the Russian River in Sonoma County, the accumulation of vegetation and debris along the river reduced the stream's capacity, raising water levels to record highs. The January storms also caused extensive flooding in the Sacramento River basin, primarily due to storm drainage system failures. The Salinas River set a new record for water levels, exceeding the previous record by four feet. Heavy development in floodplains elsewhere led to bridge collapses, streets turning into rivers, and flooded downtowns, including portions of San Jose. Total damage of the early 1995 floods was estimated at over $5 billion (in 2022 dollars).

2005 SOUTHERN CALIFORNIA

Over January 7–11, 2005, Southern California, and particularly the city of Los Angeles, received 10–20 inches of rain, essentially the entire average annual rainfall (16 inches) in just five days. This was the greatest rainfall year since 1884. Streams throughout Southern California, including those of the Los Angeles River basin, and the Ventura, Santa Ynez, and Santa Clara Rivers all crested and flooded, causing widespread damage. Seventeen people died in the event, with several hundred others displaced by flooding. The 2005 event was the first large flood in Los Angeles County since 1938 and affected communities along the Los Angeles River and areas ranging from Santa Barbara County in the northwest to Orange and San Diego Counties in the south, as well as Riverside and San Bernardino Counties to the east. Total damages were estimated at between $310 and $460 million (in 2022 dollars).

2005–06 NORTHERN CALIFORNIA

A series of storms that began before Christmas 2005 on December 17 and ended after New Year's Day 2006 produced heavy runoff over much of Northern California as well as southwest Oregon and western Nevada. While the initial dry soil conditions and time between storms of mid- to late December allowed rivers to rise and fall with little flooding, this changed with the larger storms arriving between December 28, 2005, and January 2, 2006. Several rainfall stations in the Sierra Nevada had precipitation greater than 20 inches for the December 24 to January 3 period, and this led to the Truckee River overtopping its banks and flooding Reno and Sparks, Nevada. Some stations in the Coast Ranges exceeded 18 inches during the storms, and the smaller streams, such as the Russian and Napa Rivers, that respond quickly to intense rainfall produced major flooding. Several of California's north coast rivers, including the Klamath and Eel Rivers, climbed to several feet above flood stage and nearly reached the levels that were experienced during the December 1996–January 1997 flood.

More localized flooding from the heavy rains occurred in the San Francisco Bay area and greater Sacramento area as smaller creeks and streams flowed into neighborhoods and flooded streets. The storms also created problems for the state highway system, with mudslides and rockfalls leading to road closures on Interstate 5 near the Oregon border and Interstate 80 near Truckee. The westbound lanes of Interstate 80 in Fairfield were under four feet of water, forcing the closure of the main highway between San Francisco and Sacramento. Total storm damage was estimated at $420 million.

2017 RUSSIAN RIVER AND OROVILLE DAM SPILLWAY FAILURE

Heavy rains in January 2017 caused the Russian River to climb to three feet above flood stage, inundating about 500 homes and leading to the evacuation of over 3,000 people from the Guerneville area, a site of frequent flooding. This picturesque Sonoma County community was built along the Russian River on its natural floodplain. Since 1940, the river has washed over its banks an impressive 38 times, or nearly once every other year. While a large dam has been proposed, the relatively small population in this area would not make this a cost-effective solution. What has been proposed as a more economically feasible approach is to elevate the homes and businesses above the flood levels, move them to higher elevations, or tear them down. About 250 of the town's roughly 450 flood-prone homes had been elevated by 2019, helping to reduce future flood damage.

February 2017 brought prolonged rainfall and runoff to the upper Sacramento Valley that poured into Lake Oroville, leading to a completely full reservoir and overflow down the concrete spillway. The Oroville Dam, with a height of 770 feet, is the tallest dam in the United States. It was completed in 1968 in the Sierra foothills north of Sacramento as an earth-fill dam that is owned and operated by the California Department of Water Resources. In order to handle potential overflow, the dam was built with a concrete spillway about 180 feet wide and

3,000 feet long that dropped in elevation approximately 500 feet to the river below. There was also an unlined emergency spillway with two sections, where after passing over a concrete weir, the overflow would flow down a natural hillside to the river below. The failure of a section of the concrete spillway and the potential for a more serious breach led to an extensive investigation of the dam spillway system.

While the February 7 flow down the concrete spillway was only about a third of the highest previous flow, this chute experienced failure of a large concrete slab about halfway down its length. Once the flow penetrated the concrete, it began to erode the soil beneath and along the side of the spillway (figure 5.12). While efforts were made to reduce the flow, the high runoff into the reservoir from the heavy rainfall over the watershed led to a continuing rise in water level. This led to the water overflowing the emergency spillway, which had never happened before in the dam's history. The hillside began eroding rapidly because of the large volume of water, which was actually less than four percent of what it was designed to handle.

Concern with the potential for undermining the concrete weir at the crest of the dam, which could have produced an uncontrolled release of water from the reservoir, led to the decision to increase the flow down the concrete spillway, as well as an evacuation order for about 188,000 people downstream from the dam. Officials feared the collapse of the emergency spillway, which could have sent a 30-foot wall of water into the Feather River below and flooded communities downstream. This clearly was an extremely delicate and dangerous situation that was totally unexpected.

While erosion continued beneath the concrete spillway, a large disaster was averted. This was one of the most serious dam safety incidents in US history, due to the very large size of the dam, the volume of water in the reservoir, and the downstream population. Total cost of dam and spillway repairs reached $1.1 billion, equal to the original cost of dam construction in 1967 ($1.1 billion in 2022 dollars).

The nine-month-long independent forensic analysis by a team of experts released a 584-page report with a number of critical findings. The report concluded that the spillway incident was the result of interactions of numerous human and physical factors, beginning with the project design and continuing during the subsequent half century. There was no one root cause, but rather the incident was fundamentally the result of a long-term systemic failure to recognize and properly address the deficiencies and warning signs that preceded the incident. The

FIGURE 5.12. High flow down the concrete spillway of the Oroville Dam on February 27, 2017, led to collapse of a portion of the spillway. *By Dale Kolke, courtesy of California Department of Water Resources, public domain.*

systemic failure involved practices of the dam's owner, its federal and state regulators, its consultants, and the nation's general industry practices. The incident, therefore, could not reasonably be "blamed" on a single individual, group, or organization.

Two summary statements in the report best sum the situation up: First, the dam industry in the United States lacked effective organization and dissemination of technical best practices information for dam engineering and dam safety throughout the industry, resulting in insufficient technical expertise of many engineers and geologists involved in the Oroville Dam project. These engineers and geologists "didn't know what they didn't know." Second, "although the practice of dam safety has certainly improved since the 1970s, the fact that this incident happened to the owner of the tallest dam in the United States, under regulation of a federal agency, with repeated evaluation by reputable outside consultants, in a state with a leading dam safety regulatory program, is a wake-up call for everyone involved in dam safety."[14]

DECEMBER 2022–JANUARY 2023 RAINFALL AND FLOODING

After two decades of the most severe drought in California over the past 1,200 years, a series of nine atmospheric rivers began to impact California on December 26, 2022, and continued for the next three weeks until January 17, 2023. For certain areas of the state, 2022 was the second-driest year in over 128 years of record, and the great majority of California was classed as experiencing exceptional, extreme, or severe drought (see chapter 6 on Climate Change and Drought). In what has been termed the whipsaw effect, the weather of the state changed from drought to flood within a week's time, which was confounding to both water planners and politicians. One day they were planning for a continuing drought and needing to ration nearly every gallon of water remaining in the state's reservoirs, and a week later the challenge was reversed and managers and operators were concerned about dam storage capacity, overtopping of levees, and flooding, all of which happened in late December and early January.

The rainfall intensity was intense and prolonged. Whereas single atmospheric rivers have become a common phenomenon in California, with these events typically delivering 30–50 percent of our annual precipitation, nine of these in a continuing parade was unexpected. These corridors of atmospheric water affected the entire state, from Crescent City in the north to San Diego in the south, with some areas in the central part of the state setting three-week rainfall records and some stations recording their average annual precipitation in that three-week period. Oakland set a new record for 24-hour rainfall with 4.74 inches on December 31, and across the bay, San Francisco was drenched with 5.46 inches of rain, the second-wettest day in the city's history. In South San Francisco, Highway 101 was flooded. In the Central Valley, State Route 99 was flooded when a levee on the Cosumnes River failed, trapping some people in their cars.

Evacuations were ordered for both Ventura and Santa Barbara Counties, including Montecito, which had experienced disastrous and deadly mudflows in 2018. The Santa Barbara Municipal Airport in Goleta was closed due to flooding. Rail traffic on the Santa Paula Branch Line was halted as a bridge over Sespe Creek near Fillmore was washed away.

A number of different highways were also temporarily closed in Southern California. A flash flood in a normally small creek near the town of San Miguel in San Luis Obispo County swept a five-year-old boy out of his mother's grasp as they were exiting their flooded car on the way to school. The Salinas River overtopped its banks at several locations, leading to a levee break and flooding near Salinas. Water levels beneath a major highway bridge in Paso Robles (the 13th Street bridge) climbed so high that the bridge was temporarily closed to vehicular traffic.

Streamflow in the state's rivers, starting from low-flow drought conditions in late December, tracked these nine storms clearly as the rainfall and runoff reached the rivers and moved downstream. The San Lorenzo River drains 138 square miles of mostly steep slopes in the Santa Cruz Mountains. Runoff can reach the mouth from the upper reaches of the watershed in about 12 hours, so the discharge responds quickly, and the hydrograph (record of water flow) of late December into the middle of January captured each of the atmospheric rivers. The discharge at the USGS gauging station in Felton (Big Trees) jumped from 23 cfs on December 26 to 1,570 cfs 18 hours later. By December 31 it had increased to a peak of over 13,000 cfs. If we put an impermeable wall around a football field to form a giant pool, this peak flow on the San Lorenzo River could fill that pool in 3.3 seconds.

On the Santa Clara River, one of the largest in the southern part of the state, draining 1,030 square miles of arid, mostly chaparral-covered mountains, six distinct peaks were recorded. These corresponded to the six largest atmospheric rivers (figure 5.13) producing very steep peaks, with the discharge decreasing rapidly after each storm.

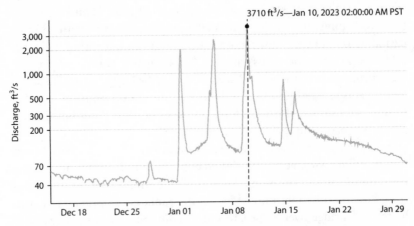

FIGURE 5.13. Hydrograph (record of water discharge) of the Santa Clara River at Piru during the series of atmospheric rivers in late December 2022 and early January 2023. The river flow (in cubic feet per second—ft³/s or cfs) went from being very low (~60 cfs) to over 2,000 cfs within a few days and rose and fell with each new storm. *Courtesy of US Geological Survey, public domain.*

Damage from flooding in 2023 was extensive to private homes, vehicles, and businesses as well as public infrastructure, including streets and highways and power and communication lines and facilities. Over 200,000 homes and businesses lost power due to the storm, at least 6,000 people were ordered to evacuate their homes, and 22 fatalities were reported across the state. The statewide storm impact led Governor Gavin Newsom to declare a state of emergency on January 4. President Biden visited storm-damaged areas in Santa Cruz County on January 19 and ultimately declared 41 of the state's 58 counties disaster areas. This declaration authorizes the Federal Emergency Management Agency (FEMA) to identify and task federal agencies to mobilize federal supplies, resources, and equipment that have been requested by the state to perform protective measures in designated counties. Over 40 California state parks were closed, as was one national park, Redwood National Park in Humboldt County, where some combination of heavy rainfall and several moderate earthquakes led to a landslide that closed Mattole Road in Humboldt Redwoods State Park.

While rainfall persisted and flooding continued in the lower elevations across California during this series of atmospheric rivers, snow fell in record amounts in the Sierra Nevada. Snowfall by mid-January was about 200 percent of normal and was wet, holding twice as much water as during an average year. The Sierra snowpack has historically been California's life vest in providing about 30 percent of the state's water and doling it out in the summer months when the river flows have dropped and the snow melts. However, with the climate changing, high temperatures mean that the snow cover will melt earlier in the year, and also that warm spring rains at higher elevations can melt the snow sooner.

The prolonged and high-intensity rainfall from late December to mid-January overwhelmed a number of creeks and rivers as well as the storm drains and runoff capacity of streets and highways, trapping many motorists and damaging large numbers of motor vehicles. Damaging flooding occurred along the Russian, San Lorenzo, Salinas, Pajaro, Carmel, and Santa Ynez Rivers, as well as many other smaller streams. From historic records of the extent of flooding in some localities or in some communities, it has become clear that because of specific geographic locations and topography, there are many neighborhoods and communities that have been repeatedly inundated. One location along the central coast is unfortunately a good example of this phenomenon, the Felton Grove neighborhood in Santa Cruz County, but there are many others.

Felton Grove is a neighborhood of 60 small homes and cabins adjacent to the San Lorenzo River about eight miles upstream from its mouth in the city of Santa Cruz. This area was flooded 14 times in a 46-year period, or about once every 3 years (see figure 5.10).[15] The 100-year floodplain is a well-known concept; it's the area adjacent to a river or stream that would flood once every 100 years on average. In comparison, Felton Grove is on the 3-year floodplain. Flooding of this low-lying neighborhood was discussed earlier, in the section on the 1982 San Francisco Bay area floods. Water levels in the 2022–23 flooding and in 1982 were comparable and led to the same type of damage, and while some homes had been elevated, high water reached four to five feet and entered garages where many appliances, tools, and other possessions were stored, and damaged or destroyed.

In many flood-prone areas, homeowners either cannot obtain or cannot afford flood insurance, which isn't normally part of a homeowner's insurance policy. FEMA has paid for repeated flood damage for decades, and the agency has been in debt for years, but these outdated policies are beginning to change, and annual premiums are starting to reflect the actual probabilities of flooding and the costs of flood damage.

SOME FINAL THOUGHTS ON FLOODING

While it takes time to document all of the damage and losses from a series of storms of the magnitude and extent of the 2023 winter, early estimates range from "over \$1 billion" to "over \$30 billion," emphasizing the uncertainties of determining these types of losses. Whatever the final number is, it will be high and the first billion-dollar US disaster of 2023.

California's residents have seen more droughts than floods in recent decades, but one lesson from the most recent 2022–23 atmospheric rivers is that we can actually go from drought to flood in a week. This is exactly what happened in the last week of December 2022, to the surprise of most people. The previous discussion of the flood history of California is a reminder that we are never too far away from another very wet winter. With a warming planet, climate models indicate, we are likely to see hotter, drier summers and to see winters with more concentrated rainfall, and this seems to be the pattern we are increasingly experiencing.

Many of the state's towns and cities were built all or in part on the historic floodplains of rivers and creeks. While levees have been built along some of these streams for flood protection and over 1,400 named dams have been constructed, usually for flood control and some combination of water supply, recreation, and hydroelectric power, all protection ends somewhere. This is just a fact of life. We cannot afford to, nor do we, build levees or dams to provide protection from the 500-year flood, for example. Yet such an event has happened and can again. The damage to the concrete spillway and then the emergency spillway at the Oroville Dam pointed that out. Engineers plan for some reasonable or agreed-upon level of protection; for levees this may be their best estimate of the 100-year event. But depending upon the number of years of discharge record for a river, determining the 100-year event may be more of an art than a science. And with climate changing, the magnitude of flooding or size of the floods on individual streams will also very likely be changing as precipitation patterns change.

In addition, as the original natural landscape has been altered by urbanization, the vegetated areas, whether forests or fields, have been partially or, in some areas, completely covered with impervious surfaces—roofs, streets, sidewalks, and parking lots. This process leads to less infiltration into the subsurface and more rapid and concentrated runoff, increasing downstream flooding. Recognizing the risks of where we have built is the first step; how we respond to those risks as individuals or communities is far more challenging, and our historic approach has been to build higher levees rather than relocate communities to safer areas. All flood protection ends somewhere, however, and to avoid future disasters, we need to learn from the floods and destruction of the past and accept that there are areas of repeated flooding that will never be safe. At the point when flood insurance rates reflect the true exposure and risks, it will simply no longer be affordable to rebuild in these areas.

6

Climate Change and Drought

Nearly everybody in California has been feeling, and in many cases suffering from, the effects of record high temperatures and the associated dry winters, prolonged drought, and record wildfires over the past two decades. Society's need to cope with and adapt to a changing climate and the associated environmental conditions isn't new, however; the stakes are just much higher today than in the past. Humans have been adjusting to their environment since the dawn of civilization roughly 8,000 years ago. Agriculture is one of the earliest examples: over the ages, farmers have repeatedly modified cultivation practices and bred new plant and animal varieties that were suited to changing climatic conditions. In recent times, dams and reservoirs, crop insurance programs, flood protection and floodplain regulations are a few examples that reflect our efforts to stabilize and protect our farms, homes, industries, and livelihoods as well as our food and water supplies in response to a varying climate.

The 2012–22 period was one of the driest on record in California history, yet we have still experienced intense and concentrated winter rainfall, runoff, and flooding, including most recently, the winter of 2023, which seriously impacted the state while this book was in final stages of editing. Precipitation in 7 of the last 10 years has been below the statewide average of 22.9 inches, and the 2012–15 *water years* (water years extend from October 1 of one year to September 30 of the

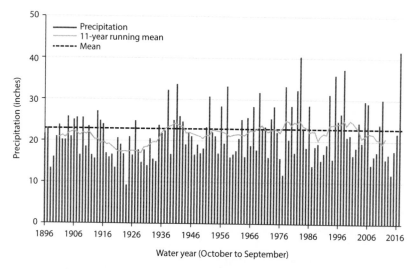

FIGURE 6.1. Long-term statewide annual mean precipitation history for California. *Courtesy of California Office of Environmental Health Hazard Assessment (OEHHA).*

next in order to capture winters in a single year) set a record for the driest consecutive 4-year period in statewide precipitation history (extending back to 1895). Yet, the 2017 water year was the wettest overall in California's history, with a state average of 41.66 inches of precipitation (figure 6.1).

In striking contrast to the natural disasters discussed in earlier chapters, however, climate change is a longer-term process. It is not an instantaneous or nearly instantaneous event like an earthquake, landslide, or flood. One way to compare the two is to think in terms of climate being what we predict and weather being what we get, or climate is long-term and weather is short-term. Although the frequency and intensity of weather-related events—like floods or severe El Niño winters with heavy rainfall and large waves—can change over decades as climate changes, we can't look at just one year or even several years and assume or proclaim that the storm we just suffered through was driven by global climate change. Climate changes over decades and is a long-term process. But we have now measured and observed countless indications that climate is changing globally, and these changes are affecting and will continue to affect California in significant and mostly negative ways. We have also made progress in what is now known as attribution science: based on our historical record of climate-related disasters—be they flooding from atmospheric rivers, wildland fires, or hurricanes—

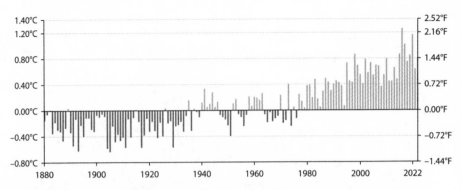

FIGURE 6.2. Global land and ocean temperature anomalies for February relative to the mean (average) of temperatures from 1900 to 2000. *Courtesy of NOAA, public domain.*

we can conclude or estimate that the individual rainstorm or flood was *x* percent greater or more severe due to the effects of an altered climate.

The continuing warming of the Earth's land and the oceans has been well documented from thousands of temperature recording stations around the planet over many years (figure 6.2). With the exception of 1998, 18 of the warmest years on record globally (since 1880) have all occurred since 2000. This trend is not just a coincidence. Scientific consensus among climate scientists, based on an overwhelming body of evidence, is that climate change is happening, it is caused in large part by human activities, and unless urgent action is taken at all levels of government to both mitigate and adapt to it, the human population and our surrounding environment could experience increasingly serious and damaging impacts in the decades ahead.

Since 1895, average annual air temperatures in California have increased by about 2.5 degrees Fahrenheit (°F) (figure 6.3). Warming has occurred at a more rapid rate since the mid-1970s. Nine of the 10 warmest years on record occurred between 2012 and 2023; 2023 was the warmest year on record to date. Temperatures at night, which are reflected as minimum temperatures, have increased almost three times more than daytime temperatures.

DROUGHT IN CALIFORNIA

There is no shortage of natural disasters to contend with in the state, but compared to the calamities discussed in other chapters—earthquakes, floods, debris flows, and El Niños—droughts are subtle,

Degrees Fahrenheit

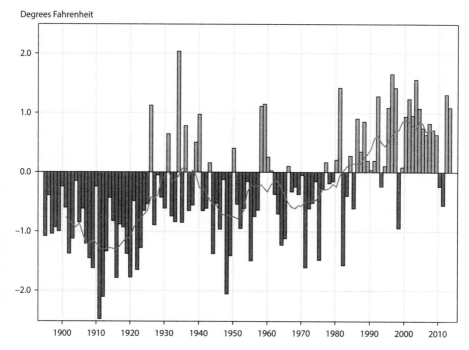

FIGURE 6.3. Temperature history for California from 1895 to 2014 using departure from the mean for the 1949–2005 period in degrees Fahrenheit. The gray curved line is the 11-year running mean. *Courtesy of Western Regional Climate Center/NOAA, public domain.*

and much more difficult to evaluate, document, and quantify. Unlike an earthquake, a drought is not instantaneous, and the effects are spread out over a much longer period of time. Drought is a gradual phenomenon, and the impacts are usually felt first by those most dependent on annual rainfall—ranchers engaged in dryland grazing, farmers, rural residents relying on wells in low-yield aquifers, or small water systems lacking a reliable water source. Drought impacts will increase with the duration of a drought, as reservoir levels decline and groundwater levels in aquifers drop.

California now has about 150 years of rainfall history, and throughout much of the last century, chambers of commerce and the real estate profession often claimed that the state had the greatest climate in the world. When severe floods occurred or the rains missed California for months on end, these years were often described as "exceptional" or "unusual." Before the construction of large reservoirs, Central and Southern California were usually equally subjected to floods and droughts. Some of the oldest newspaper accounts documented the floods of 1815–25, followed by the drought of 1827–29, and then the floods of 1832 and 1842.

The 1856–57 drought followed 20 years of near-normal or above-normal precipitation, which had led to overstocking the luxuriant pastures. During the summer of that year and the subsequent winter, an estimated 100,000 cattle were lost in Los Angeles County alone. Following the legendary floods of 1861–62 came the drought of 1862–64, which may have been the driest of all historic droughts in Southern California. As will be described in later pages, these dry years more or less brought about the end of beef cattle as the state's primary industry.

What constitutes a drought? In some cases we can't even agree on this. One simple definition of *drought* is "a prolonged period of abnormally low rainfall." And a new term has been added to our weather conversations in recent years, *megadrought,* defined as "a prolonged drought lasting two decades or longer." This is a word most water suppliers would rather not think about, but evidence for such events is clear in the prehistoric records of tree rings.

When does a drought end? For the state as a whole, about one-third of our water comes from the Sierra snowpack, another third from groundwater, and the rest from surface runoff, much of which is captured in reservoirs. So after a year or two or more of below-average precipitation, we need enough rain and snow falling in the right places to bring the snowpack up, refill the reservoirs, and recharge the aquifers—which can take several years or longer—in order to end a drought.

There are difficulties in describing and comparing historic droughts with modern dry spells such as the 2012–16 period that the state broke out of—at least temporarily—in 2016–17 (see figure 6.1). The demographic, population, and economic conditions are very different today than they were 50, 100, or 150 years ago, however. California's population has grown from 380,000 in 1860, to 1.5 million in 1900, 10.6 million in 1950, 20 million in 1970, 30 million in 1990, and about 39 million in 2023. The economic base has shifted on the agricultural side, from cattle grazing, which dominated until the latter part of the 1800s, to California becoming the largest producer of fresh fruits, vegetables, and nuts in the nation. The growing population with their increasing demands for water, and the development of a water-intensive agricultural base—in contrast to cattle grazing on open ranges—are just two examples of changing water demands. Several dry years that may not have been considered a drought a century ago, when the state had just

3 million people, present vastly different challenges today with a population of nearly 40 million and 7.5 million acres of crops dependent on regular irrigation.

HISTORIC DROUGHTS AND MEGADROUGHTS

Not surprising to longtime California residents, droughts are not new or even uncommon. We get these dry periods every decade or two. If you really want to get concerned and lose sleep, the prehistoric record has preserved far more serious droughts in the distant past than we have seen over the past 150 or so years of recordkeeping in California, *megadroughts* in today's terminology. But this was before California was home to 39 million people and our farms and fields were providing fruits, vegetables, nuts, and livestock to the entire nation. In the early years of the last century, if there wasn't enough water where we wanted a city or farm, we built dams, reservoirs, canals, and pipelines to move the state's water to where it was deemed to be needed. And more often than not, we fought over who had the rights to the water. In the "Owens Valley water wars" of 1924 and 1927, farmers protesting Los Angeles stealing their Owens Valley water actually dynamited aqueducts. Nonetheless, between 1860 and 2000, 1,400 larger dams (over 25 feet high) were built on California's rivers and streams; that's an average of 10 per year or almost a dam every month for 140 years.

Droughts over the past two centuries in California have usually lasted several years (see figure 6.1), and there are numerous historic accounts of how the dry periods in the 1800s affected livestock, agriculture, and the people living here at the time. Until the last decade or so of the 1800s, however, there were no reservoirs to store winter runoff, so low rainfall years hit the early residents particularly hard. An important question to try and answer as we look towards the future is whether the last 150–200 years of climate and rainfall were typical.

Rainfall records in the state only go back about 140 years, but *dendrochronology*, or the study of tree rings, has allowed us to look much further back in time. Just like putting on a few extra pounds around your waist when you eat well on a vacation or over the holidays, trees suck up moisture when there is plenty to go around and use that extra water to grow thicker growth rings.

History has been written and recorded in many places other than in books, and the job of a paleoclimatologist is to find where historic climate records have been preserved. Tree rings, lake and seafloor sediments, corals, and ice are a few places where we have been successful in extracting long-term climate records.[1]

Bristlecone pines, which can live to be 4,000–5,000 years old and survive in the White Mountains of southeastern California, are living history books in which the records of our prehistoric rainfall have been preserved. These ancient trees contain evidence in the widths of their annual rings that the last few thousand years have been characterized by alternating 50- to 90-year wetter and drier periods, but also by droughts that have lasted 10–20 years. These are mild events, however, compared to the period from about 900 to 1400 AD, known globally as the medieval warm period. Evidence from detailed tree ring studies indicates that droughts as long as a century were common during this period.

Most climate scientists would agree that the past century was unusually wet, and it has been during the past 100 years that the populations of the arid southwestern states literally exploded. Nevada went from 42,000 people in 1900 to 3.2 million in 2023. Arizona grew from 123,000 to 7.4 million during the same period. And these states are both deserts and have a lot of thirsty golf courses.

California has 39 million people today, over 25 times more than in 1900. And all this growth, whether farms, factories, or cities, took place during what was very likely an unusually wet century. For decades, Central Valley agriculture had everything—sunshine, fertile soils, and water. But that water was imported from elsewhere or pumped from aquifers that were being depleted year after year. In 2015, the fourth year of drought, 560,000 acres of productive farmland in California were left fallow due to the drought. In 2022, as the drought continued, California's irrigated farmland shrank by 752,000 acres. Normally productive farmland went unplanted as Central Valley farmers saw water deliveries cut by nearly 43 percent in both 2021 and 2022. As of November of 2022, all 58 of the state's counties were designated as being under some level of drought, with the vast Central Valley in either extreme or exceptional drought. The prolonged rainstorms (atmospheric rivers) of January of 2023, however, had a major impact on this extended drought. All areas that had been classed as exceptional or extreme drought areas were reclassified as either severe or moderate drought areas (figure 6.4).

These almost continuous rains of the first week of January 2023 also had a significant impact on the state's major reservoirs. On November 3, 2022, California's six largest reservoirs were on average at just 30 percent of their total capacity, and 55 percent of their historic average. Lake Oroville (figure 6.5) and Shasta Lake, the state's two largest

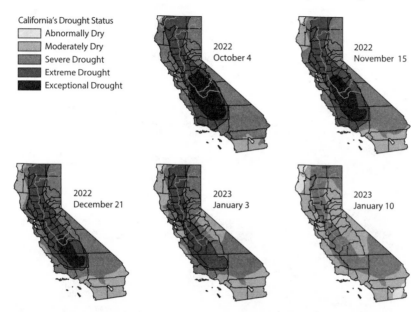

California's Drought Status
- Abnormally Dry
- Moderately Dry
- Severe Drought
- Extreme Drought
- Exceptional Drought

2022 October 4

2022 November 15

2022 December 21

2023 January 3

2023 January 10

FIGURE 6.4. October 2022 through January 2023 drought status for California's different regions as the state received the multiple atmospheric rivers of the 2022–23 winter. *The US Drought Monitor is jointly produced by the National Drought Mitigation Center (NDMC) at the University of Nebraska–Lincoln, the United States Department of Agriculture, and the National Oceanic and Atmospheric Administration. Map courtesy of NDMC.*

reservoirs, with capacities of 3,537 acre-feet and 4,552 acre-feet, respectively, have been well below their average levels for several years. Folsom Lake, while considerably smaller, has been a favorite location for houseboating for years, and the drought has significantly curtailed that activity, with major marinas now on dry land (figure 6.6).

If the state's present dry status is just the early stages of an even longer-term drought, the impacts and implications for California are enormous. It's hard for many to believe this is possible, but history tells us it has happened in the past. Most of the state's residents have lived virtually their entire lives in times of abundant water, but we need to begin coming to grips with the potential of a different future. This became evident in 2022 when water levels in both Lake Mead and Lake Powell on the Colorado River dropped to their lowest levels since Hoover and Glen Canyon Dams were built in 1936 and 1966, respectively.

The Colorado River is a critical source of water for 40 million people, a provider of hydropower, a recreation area, and habitat for fish and wildlife, affecting Arizona, California, Colorado, Nevada,

FIGURE 6.5. Lake Oroville Reservoir on October 28, 2021. *By Andrew Innerarity, courtesy of California Department of Water Resources, public domain.*

FIGURE 6.6. Boat slips at Folsom Lake on dry land on July 28, 2021. *By Ken James, courtesy of California Department of Water Resources, public domain.*

New Mexico, Utah, and Wyoming as well as parts of northern Mexico (figure 6.7). The river's water was allocated a century ago in November 1922 when delegates from each of these states met and developed the Colorado River Compact, which apportioned the river's water between what were designated as upper and lower basin states. One of the driving forces for the compact was the concern with California's growth and its increasing water demand, which was made more galling by the fact that California contributed the least runoff to the river. In May 1997, a symposium was held in Santa Fe, New Mexico, on the 75th anniversary of the signing of the compact. The general manager of the Southern Nevada Water Authority commented during her presentation that "things have changed, but what remains the same is that California was the problem back then, and California is the problem today."[2]

The original compact between the states was developed in large part to avoid federal intervention and to head off potential litigation between the states. The main accomplishment was to apportion Colorado River water equally between the upper basin states (Wyoming, Colorado, Utah, and New Mexico) and lower basin states (California, Arizona, and Nevada), with each basin receiving 7.5 million acre-feet (maf) of

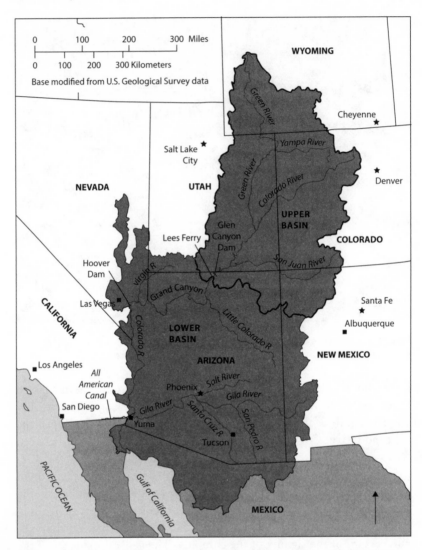

FIGURE 6.7. The Colorado River watershed and the aqueducts and canals that move that water around. *Courtesy of US Geological Survey, public domain.*

river water per year. These allocations were based at the time on hydrologic data from the Bureau of Reclamation that indicated the average annual flow of the river at Lees Ferry was 16.4 maf. This may have been a reasonable estimate at that time, a century ago now, but in reality, the average flow was more like 13.5 maf, although flows are highly erratic, ranging from about 4.4 to over 22 maf.

So not only were allocations based on overly optimistic flows, but we are now in the midst of the most severe 20-year drought in recent history. The water just isn't there to allocate, and the states are trying to agree on how to appropriate the water that is there, in order to keep the federal government from intervening. With a warming climate and a historic prolonged drought, California is going to have to either significantly reduce its usage of Colorado River water or find other sources, which is easier said than done. California used up all of the readily available and easy-to-access water resources many years ago.

On a positive note, at least in the short term, the runoff from the continuous rainstorms during the first two weeks of January 2023 had a significant impact on the state's reservoirs. California's six largest reservoirs on June 15, 2023, averaged 86 percent of their *total capacity,* up from 30 percent in November 2022; they averaged 119 percent of *average capacity* for this time of year, which was up from 55 percent in November. Storage volume can be substantially augmented with a few major rainstorms, or large atmospheric rivers. Only time will tell, however, whether these storms will have a significant impact on water availability in the following summer and fall months. Can two weeks of rain end a major drought?

The Drought of 1862–64 (the Great Drought)

There are numerous records or accounts of the drought history of the American Southwest, including California, which list specific dry periods and often their impacts. There seems to be some general agreement among these sources that the drought of 1862–64 had some of the greatest impacts of any of the 19th and 20th centuries, which somewhat oddly, came immediately after the greatest flood in the state's history in 1862.

Until this severe dry period, a major part of the young state's economy and even its way of life, primarily that of the *Californios,* was tied directly to cattle ranching. *Californio* is a Spanish term generally defined as "persons of Spanish or Mexican heritage whose place of birth or residence was California." In most cases these were the early residents who were given the major Mexican land grants, many of which eventually became the large cattle ranches. And many place names in California were named after the original *Californios*—Pico, Vallejo, Alvarado, Castro, Pacheco, Carrillo, and many others. In the words of California historian Sandy Lydon, the drought of 1862–64 was incredibly hard on the Californios who were trying to hang on to their pastoral culture. Many California historians have written that the 1862–1864 drought ended the Californio culture forever. . . . Strapped for cash, the Californios went to the money-lenders and put the only thing they owned—their land—up for collateral. It had never stayed dry for longer than one season, they reasoned. Surely it would rain in December or January. But the storms stayed north and the next spring and summer the sound of the auctioneer's hammer echoed above the bawling of the dying livestock. Some ranchos fell for 10 cents an acre. David Jacks bought the 8,794-acre Los Coches Rancho [Monterey County] for $3,535. The names Garcia, Soberanes and Castro were removed from deeds and Spence, Jacks, Iverson and Hihn put in their place.[3]

In an excerpt from "Exceptional Years: A History of California Floods and Droughts," J.M. Guinn in 1890 wrote, apparently describing the rainfall in Southern California:

1862–63 did not exceed four inches, and that of 1863–64 was even less. In the fall of 1863 a few showers fell, but not enough to start the grass. No more fell until March. The cattle were dying of starvation. . . . The loss of cattle was fearful. The plains were strewn with their carcasses. In marshy places and around the cienegas [wet meadows], where there was a vestige of green, the ground was covered with their skeletons, and the traveler for years afterward was often startled by coming suddenly on a veritable Golgotha—a place of skulls—the long horns standing out in defiant attitude, as if protecting fleshless bones.[4]

The Extended Drought of 1929–34

The 1929–34 drought occurred during the infamous Dust Bowl period that impacted the Great Plains of the United States in the 1920s and 1930s and brought many new immigrants to California. It occurred within the context of a decades-plus dry period in the 1920s–1930s (see figure 6.1), the driest in the past century, and a period whose hydrology rivaled that of the most severe dry periods in more than a thousand years of reconstructed California Central Valley paleoclimatic data. The drought's impacts were small by present-day standards, however, since the state's urban and agricultural development, and therefore water demands, were far less than those of the 21st century. The state's population in 1930 was just 5.6 million, compared to 39 million in 2023, a sevenfold (700 percent) increase. In response to this drought, the vast Central Valley Project designed to move water around California—from where it was to where it wasn't—was begun in the 1930s.

Sacramento has one of the longest precipitation records of any location in California. While the long-term average annual rainfall (1850–2014) is 18.34 inches, the 30-year moving average rainfall varies from 20.42 inches in 1896 down to 14.51 inches in 1937 and then up again to 20.47 inches by 2007. This is a 30–40 percent swing in 30-year average rainfall values within a single lifetime. The 1929–34 drought marks the low point of the extended dry period that spanned the 1920s and 1930s.

The Drought of 1976–77

While the drought was short in duration, 1976–77 was one of California's driest two-year periods on record (see figure 6.1). Statewide annual precipitation was just 15.8 inches in 1976 and 11.6 inches in 1977, compared to a long-term mean of 23 inches. Precipitation for most parts of the state was less than half of normal, with some areas receiving just 15 percent of average. These two dry years served to wake up many of California's water agencies to the fact that they weren't prepared for major reductions in their supplies. One important reason for the lack of preparedness was the long-held perception that the state had always had ample water supplies. In addition, there had not been major droughts in the recent past, and people in general have both short disaster memories and also short drought memories. Although there had been multiyear dry periods of statewide extent in 1947–50 and 1959–61, the hydrology of these years was far less severe than that of the 1920s–1930s. The impacts of the dry 1976 year were mostly mitigated by reservoir storage and groundwater availability. But with the even drier 1977 year, impacts were felt widely. River flows declined precipitously, and reservoir levels dropped dramatically in response. Forty-seven of California's 58 counties declared local drought-related emergencies with mandatory water rationing.

Along the central coast, for the first time, the city of Santa Cruz implemented water use restrictions, also known as water rationing, which is never popular. The city recorded just 13.88 inches of rainfall in 1976 and 15.93 inches in 1977, the two-year total being equivalent to the city's average rainfall for a single year. Monterey, with an average annual precipitation of 19.31 inches, had similar low rainfall, with just 9.76 inches falling in 1976 and 10.46 inches in 1977, also nearly equivalent to an average single year. Salinas, with an average annual rainfall of 13.03 inches, echoed the pattern, receiving just 6.83 inches in 1976 and 8.02 inches in 1977.

Reservoir depletion had a major impact during this two-year period. On October 1, 1976, storage in California's major reservoirs was 57 percent of average, and a year later this had declined to 37 percent. Marin County was the urbanized area most affected by the drought, with most communities in the southern part of the county being restricted to basic health and safety water consumption levels. The area has limited groundwater resources and had only local surface water supplies at the time, which has since been remedied. A temporary emergency pipeline was actually constructed across the San Rafael Bridge during that dry period to bring water to Marin from the Metropolitan Water District.

The Drought of 1987–92

California endured one of its longest extended droughts since the 1920s and 1930s (the Dust Bowl years) in the six years from 1987 through 1992 (see figure 6.1). Yearly rainfall over much of the state was just over half of the 20th-century average. While reservoirs can buffer the lack of rainfall for a year or two, when a drought lasts five or six years, trouble and shortages begin, which lead to restrictions and rationing. As a result, the users served by most of the state's larger water agencies didn't begin to experience shortages until the third or fourth years of the drought, due to deliveries from the larger reservoirs. Reservoir storage was down to about 40 percent of average by the third year of the drought, and it didn't return to average levels until 1994, as a result of a wet 1993. These longer dry periods impact everything from agricultural needs to hydroelectric power production, from loss of fish populations to declines in groundwater tables as aquifers are pumped more heavily.

By 1990, California's population had reached about 30 million, 77 percent of present levels. The state's irrigated acreage was essentially the same as today's. California was continuing to get more than its basic share of Colorado River water as a result of unused allocations of Nevada and Arizona. This helped to alleviate the State Water Project cutbacks in Southern California. The Central Valley Project (CVP) and the State Water Project (SWP) met demands during the first four years of the drought, but as reservoir levels continued to fall, they were forced to substantially reduce deliveries. In 1991 the SWP cut off deliveries to agricultural contractors and provided just 30 percent of urban deliveries. The CVP provided only 25 percent to agricultural agencies and 25–50 percent to urban contractors.

The agricultural sector didn't plant an estimated 500,000 acres, about five percent of the 1988-level harvested acreage. Estimated gross revenue losses to farms from this fallow land were about $220 million in 1990 and $250 million in 1991. The commodities with the greatest impacts were grains, nonirrigated hay, and beef cattle, with the biggest losses being on the west side of the San Joaquin Valley.

Local urban water agencies generally minimized cuts to commercial and industrial users in the interests of avoiding potential job losses and shifted the burden of water use restrictions to residential customers. There was widespread damage to timber resources throughout the Sierra Nevada due to infestation of bark beetles. The extended duration of this drought set the stage for a pattern that would emerge in subsequent extended dry periods, namely the connections between severe droughts and the risk of major wildfire damage in densely populated urban areas and at the wildland-urban interface. The six-year drought in California ended in late 1992 as a major El Niño event in the Pacific most likely led to the unusually persistent heavy rains.

Droughts of 2007–09 and 2012–14

The 2007–09 period was ranked as the seventh-worst three-year drought period in the state's recorded rainfall history (see figure 6.1), and it was the first for which a statewide proclamation of emergency was issued. This dry period also saw greatly reduced water diversions from the SWP and in some ways was another wake-up call for what was to follow a few years later. Drought impacts were again most severe on the west side of the San Joaquin Valley where CVP deliveries in 2009 were only 10 percent of the farmers' allocations, following deliveries of 50 percent in 2007 and 40 percent in 2008. These unprecedented reductions in CVP and SWP diversions from the Sacramento–San Joaquin Delta had very significant economic impacts on agriculture and on those rural communities that depend on agriculture for employment. They were exacerbated by the diversion from the delta to protected endangered fish species, which was a relatively new regulatory requirement. The summer of 2007 also saw some of the worst wildfires in Southern California history.

The 2012–14 water years became the driest three-year period in the long history of measured state rainfall. The allocations from the CVP and the SWP reached record lows in 2014. New Endangered Species Act (ESA) listings and management of fish populations began to impact the operations of many of the state's water projects, including the large projects affected by listing of Central Valley fish species as well as smaller projects on coastal rivers. Not only have the state's water budget and the combined impacts of climate change and extended drought impacted water availability, but changes in streamflow needed to support endangered fish populations continue to create additional challenges.

The Great Drought of 2012–16

The 2012–16 drought is considered by many to be the worst in California history and was one of extreme proportions (see figure 6.1), with record-high temperatures and record-low levels of snowpack and rainfall, as well as very low river runoff, significantly below the 30-year average for the 2014, 2015, 2018, 2020, and 2021 water years (figure 6.8).[5] The five-year duration of the dry conditions, the growing population with its increasing usage, and the greater demand for agricultural water all served to make this a very serious event for the state, but one that may likely become more common in the future as the climate continues to change. Continued global warming is expected to lead to (1) less precipitation falling as snow at higher elevations, which means less snowpack for spring and summer runoff; (2) rainfall more concentrated in the winter months, which translates into more frequent and severe flooding; (3) inability to trap and store as much runoff in reservoirs because of concentrated winter runoff; and (4) warmer and drier summers, which would lead to greater water demands for urban and agricultural users, and also more brush and forest fires, as were experienced in the fall of 2017 and 2018.

By the second year of this drought (2013), California overall received less than 34 percent of average expected precipitation. For many regions of the state, the rainfall totals for 2013 were the lowest on record. Salinas, for example, got just 3.25 inches of rain the entire year (25 percent of normal), the lowest in its 83 years of recordkeeping.

As the drought continued, there were a number of direct effects. Streamflows were so low that fish couldn't get upstream to their spawning grounds. Many river mouths were blocked with sandbars that prevented salmon from even starting their spawning journey upstream. The California Department of Fish and Wildlife estimated that 95 percent of the winter-run salmon didn't survive 2013. An additional impact of the drought and extreme heat was an unprecedented die-off of trees, which increased the risk of wildfires. In November of 2016, the US

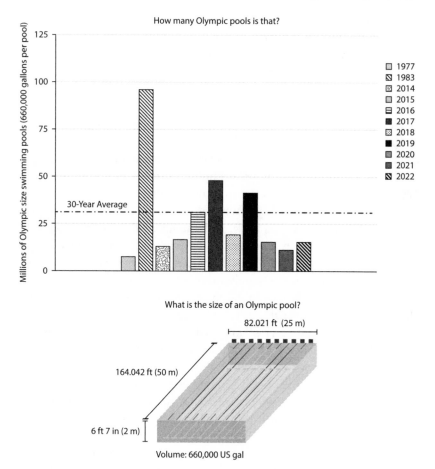

FIGURE 6.8. Annual California river runoff for recent dry years compared to the very wet 1983 winter. Runoff is expressed in number of Olympic size swimming pools (660,000 gallons) for comparison. *Courtesy of California Water Science Center—US Geological Survey, public domain.*

Forest Service reported that the extended dry conditions led to the death of over 100 million of the state's trees.

In February 2014, continuing drought conditions led the California Department of Water Resources to develop a plan to reduce water allocations to farmland by 50 percent. This was a huge issue for the state's $45 billion agricultural industry, which grows nearly half of the nation's fruits, vegetables, and nuts. By mid-May of 2014, the entire state was under a severe drought or a higher-level (exceptional drought) condition.

Dendrochronology, or tree ring studies, in the Sierra Nevada by several different groups concluded that the rainfall in California had been abnormally high since about 1600 and that the state had experienced its wettest period in 2000 years during the 20th century. This climate history does not bode well for California's future and our water dependence.

In June 2015, the governor ordered all cities and towns to reduce water usage by 25 percent, which is important, but with agriculture in California using about 80 percent of its water, domestic use reduction won't solve the state's water deficiencies. As agricultural water allocations were reduced, pumping of groundwater from wells was increased, which led to continuing decline in Central Valley water tables.

The Sierra snowpack levels were at or near record lows during the winters of 2013, 2014, and 2015, with the statewide snowpack on April 1, 2015, holding only five percent of the average for that date, a record that extends back to 1879. The snowpack normally provides about one-third of California's water supply, so this was bad news indeed.

It has often been the case in the past in the state that a very wet winter follows a period of drought, or just the opposite happens, a drought follows a very wet winter. This happened with the floods of 1861, which were followed by the drought of 1862–64. The winter of 2016–17 turned out to be the wettest on record in Northern California, surpassing the previous record set in 1982–83. Runoff filled Oroville Reservoir (the second largest in California, with a 770-foot-high dam, the tallest dam in the United States), and overflow down the spillway in early February led to partial failure of the concrete structure (see figure 5.12). This prompted the temporary evacuation of about 180,000 people downstream of the dam north of Sacramento. This near disaster required a $1.1 billion repair project. In response to the heavy precipitation, which flooded multiple rivers and filled most of the state's major reservoirs, Governor Brown declared an official end to the drought on April 7, 2017.

The 2016–17 winter rains did, however, fill or partially fill many of the state's reservoirs that had been so depleted during the previous dry years. Another reversal in hydrologic conditions followed that very wet winter. Progressing into 2018, however, both rainfall and snowpack in the state were significantly below average, bringing to mind the drought that had just ended. During the 2017–18 winter (2018 water year), the Los Angeles International Airport rainfall station, which has records going back to 1944, recorded just 3.87 inches of rainfall, the second driest in 78 years of record and just 30 percent of average. Pasadena's rainfall history extends back 116 years, and 2017–18 was tied for the third-driest year with just 8.04 inches; 2017 was the second driest in Santa Cruz history. Southern California was hit much harder by the dry water year, where most stations received between 27 and 39 percent of normal, whereas totals in Central California were generally in the range of 46 to 66 percent of average, and Northern California stations received 45 to 94 percent of normal.

SOME FINAL THOUGHTS ON DROUGHTS

The continuing warming of the planet has raised the average summer temperatures in California by about 3°F (1.8°C) since 1896, with over half of that happening since the early 1970s. If our global greenhouse gas emissions continue at the current rate—and they are still increasing—California could be over 2°F warmer by 2040, more than 4°F warmer by 2070, and over 5°F warmer by 2100. To put it mildly, this would be very bad news for the state in all respects. The increasingly frequent reoccurrence of extreme wet periods followed by dry periods (sometimes referred to as a whipsaw effect) is consistent with climate models and indicates dry years will become drier and wet years will become wetter, which will intensify both our droughts and our floods. This makes planning for future climate change ever more difficult. It also complicates decision-making as politicians at all levels are most often influenced in their decisions by what happened the previous month or year, rather than the longer-term trends. One wet year often leads to our forgetting or minimizing the prior 5 or 10 years of drought.

Climatologists in recent years have recognized narrow bands of high-moisture air arriving at the coast from the Pacific Ocean to the west and have designated these as atmospheric rivers (see figure 5.1). These now are being appreciated for both their positive and negative impacts—they deliver 30–50 percent of the state's rain and snow that supports our farms, factories, communities, and cities but also bring increasing flood, debris flow, and landslide risks, as the continuous arrival of atmospheric rivers in the first two weeks of January 2023 did. Scientists believe that these atmospheric rivers are likely to become more frequent and intense with continued global warming, which will increase the risks of flooding and slope failures.

Our mountain snowpacks provide about one-third of the state's water supply. Observations show, however, that a warming climate is negatively affecting this water source. More precipitation is falling as rain, therefore reducing the summer runoff that has historically been stored in the snowpack. Fall and spring are becoming longer, so winter is getting shorter, again reducing the amount of snowfall and therefore snowpack and subsequent runoff. Water from the Sierra snowpack could be reduced to less than a third of historical levels by 2100.

A changing climate is already affecting all Californians, and these impacts are virtually certain to get worse long before they get better. The water supply needed to maintain the state's immense agricultural industry, which is critical not only to California, but to the entire nation that depends upon the fruits, vegetables, and nuts grown in the state, is no longer certain and predictable. Farmers can no longer comfortably rely on the irrigation water they have historically received. Already large areas of productive farmland have gone unplanted in recent years due to lack of water.

We can get by without water for about three days if we are just trying to survive, but we all have a desire to do more than merely survive. The paleoclimate history of the Southwest, where bristlecone pines have left us with a record of decades-long droughts in their growth rings, provides some clues as to what the state has experienced in the past. The big difference is that we aren't a small group of Native American people who have learned how to live off the land and can relocate our villages when necessary. Today we are a state with 39 million people, which supplies about half of all of the fresh produce for the entire country, and it all requires water, large volumes of water. Taking almonds as one example, California produces 100 percent of the nation's and 80 percent of the global supply, but they are a notoriously thirsty crop; one ounce of almonds requires about 23 gallons of water. California also produces 90 percent of all the nation's wine, and grapes also require water. While we have built dams, reservoirs, canals, aqueducts, and pipelines to move water from where it is to those drier places where it isn't, these vast engineering structures still require occasional precipitation. And long-term droughts can render all of these water storage and transport structures nearly irrelevant, which is what has happened to the huge reservoirs on the Colorado River, Lake Mead and Lake Powell, in recent years.

There are only a few remedies to these long-term water shortages, and they all cost money and have their impacts. One frequently aired proposal is to build more dams to store winter runoff. The last large dam in California was completed in 1980, and essentially all of the feasible sites for large dams in California have been now utilized or developed. It seems highly doubtful that any additional large dams will be constructed in the state. There have been plans put forward to raise the heights of some existing dams in order to be able to store more water, but studies indicate increased seismic safety issues, concerns about loss of wildlife, issues with sacred land of Indigenous peoples, and very high costs.

Desalination is a proven technology for producing fresh water from the ocean, which is where 97.5 percent of the planet's water resides. In 2015, California opened the largest and most technologically advanced desalination plant in the United States in northern San Diego County. It provides 50 million gallons per day, using reverse osmosis, to water-short San Diego County. There are several smaller desalination plants along the state's coast, but there are still concerns commonly raised, including costs, energy uses, and ocean impacts. The ocean impacts are very local and can be resolved or mitigated; costs and energy usage, however, will always be issues. But any new source of water will be expensive and require considerable energy.

There are projects underway for more conjunctive use of water, using aquifers to store extra water from winter runoff that can be injected into the subsurface for temporary storage. Aquifers have many advantages over surface reservoirs: they have no construction costs, they don't fill up with sediment, there are no evaporation losses, and they don't take up valuable surface land. However, the permeability and porosity vary widely in the subsurface, and pumping surface water into aquifers isn't like filling up a surface reservoir in terms of flow rates or volumes. Water quality can be an issue, and it also takes power to pump water from deep beneath the surface.

Drought will continue to occur, and historic trends and climate models suggest that they may become more frequent and last longer in the future, which provides strong incentives for both working to reduce our water demand and considering all possible additional sources of future water.

7

Wildland Fires

INTRODUCTION

It was a very warm and dry summer day in August of 1951 when I first experienced a forest fire up close. My family had decided to leave an increasingly crowded suburb of Los Angeles and try life on a 50-acre ranch in southern Oregon between the very small towns of Wilderville and Wonder. It's an exaggeration to call either of these road stops towns, however. Wilderville in 1950 was a little general store, a gas pump, and a very small post office that comfortably held three people. I think Wonder may have been even smaller; it was Wilderville without the post office and gas pump.

For my two brothers and me, this was pretty close to paradise, with a diverse menagerie of farm animals, fields and forests, and a creek running 30 feet from our back door. We were a bit isolated, however, with no telephone, and my dad took our only car to work every day.

On that warm day, my two brothers and I and a friend who lived some distance away were down in the now-dry creek bed smoking some small sticks. These were somewhat hollow, so we could light one end and pretend we were smoking, which was a common adult pastime in those days. As we climbed up out of the creek, the neighbor kid dropped a burning stick on the ground, which was covered with dry—and as it turned out—extremely flammable grass. Within literally seconds, the dry grass had ignited and that fire was spreading very fast. We tried stomping on it, but to no avail. It quickly had consumed an area of about 100 square feet and was already way out of our control. We immediately felt a sense of hopelessness and ran the quarter mile to the log cabin that was home. We could see the smoke rising from the burning grass as the fire burned towards the surrounding trees. My mother told my older brother, who was then 10, to run up the valley to the only neighbor we knew with a phone, who was perhaps a half mile away. Trucks and crews from the Oregon Department of Forestry took a while to arrive, and it took hours to bring the fire under control. It started up again after the firefighters had left, leaving our family to fill up a trailer with buckets of water to try to extinguish the resurging flames. Although I was only eight at the time, that experience remains with me over 70 years later as a vivid reminder of how easily a fire can start and how fast it can spread under the right conditions.

Wildland fires in California have become more frequent in recent decades and have also covered increasingly larger areas. Twelve of the 15 largest wildland fires in the state's history have occurred since 2000. More than 2.7 million Californians now live in areas classified as "very high fire hazard severity zones." Several factors are contributing to this increasing exposure: more and more people are relocating to urban fringes or mountainous areas surrounded by forest- or brush-covered hillsides where fire hazards are high, and as the planet continues to warm, summers are getting hotter and drier, which provides the fuel and the optimal conditions for brush and forest fires.

While one could ask whether or not fires should be considered natural disasters, in terms of property and lives lost, they rank right up there with the other hazards we usually consider. Over the last century, 316 lives have been lost directly or indirectly from wildland fires in California, exceeding the combined fatalities from tsunamis, volcanic eruptions, landslides, and coastal storms over the same time period. The 20 most destructive wildland fires in the state have all taken place since 1991 and have destroyed a total of 40,900 structures, which have primarily been homes. This exceeds losses from all other natural hazards except earthquakes and perhaps flooding, where accurate records aren't readily available. This history and these losses make a convincing case for including wildland fires in our list of natural disasters.

Wildland fires have been ignited by lightning strikes for centuries, which would make them natural events. California also has frequent dry weather from early summer through late fall months that provide situations conducive to fires under natural conditions. In the years prior to 1800, when far more of the state's landscape was forested, an estimated 4.5–12 million acres (7,000–18,700 square miles) burned annually on average, 4.5–12 percent of the state's entire area. I do need to question the accuracy of these numbers simply because prior to 1800 there wasn't a lot of recordkeeping regarding areas burned by fires.

Some of the fires were doubtless due to lightning, but it is likely that far more were set intentionally by the native Californians, who regularly carried out burns that cleared out the underbrush and dead wood, thereby allowing new grasses and other plants to grow and also removing the accumulated fuel, which prevented large wildfires from occurring.

In stark contrast to the approach followed for centuries by the native Californians—using fires constructively—the US Forest Service for decades prioritized the practice of fire suppression. A series of large fires that burned over 3 million acres in Montana, Idaho, and Washington in just two days in 1910 had a major impact on national wildland fire policies. These fires, combined with the arguments made by several early conservationists that forest fires threatened future timber supplies, convinced local and national forest administrators, as well as the public and members of Congress, that the devastation could have been prevented if they had enough equipment and men on hand. The argument was made that only total fire suppression would be able to prevent such damaging fires in the future. A policy of suppression was gradually instituted that had two goals: preventing fires and containing a fire as quickly as possible once it started.

In order to prevent fires, the US Forest Service came out against the practice of even limited burning, even though many farmers, ranchers, and even timbermen supported it because it improved land conditions. In those early decades of the 1900s, however, the forestry profession didn't have a clear understanding or appreciation of the ecological importance of fire. The Forest Service argued strongly that any fire in the woods was undesirable because it destroyed timber. This position led to an education program designed to convince the public that fire prevention was an important goal. This even led to the emergence of Smokey Bear as a character to convey the "Only you can prevent forest fires" message.

The Forest Service also set out to develop a systematic approach to fire protection, which ultimately involved the construction of lookout towers, ranger stations, communication systems, and a network of access roads. In 1911, with the support of the Forest Service, the Weeks Act was passed; it established a coordinating structure between the federal government and the states that enabled cooperative firefighting.

Through financial incentives to states to fight fires, the US Forest Service soon dominated and led what amounted to a national fire policy.

Several severe fire seasons in the early 1930s gave an even greater sense of urgency to the policy of fire suppression. In 1933, in the midst of the Great Depression, the federal government established the Civilian Conservation Corps, which put thousands of unemployed men in the field, building firebreaks and fighting fires. Two years later in 1935, the Forest Service established what became known as the "10 a.m. policy," which declared that every forest fire should be suppressed by 10 a.m. on the day following its initial report. This, however, as we now know all too well, simply isn't possible in most cases. This policy was accepted by the other federal land management agencies, however, and fire containment efforts expanded. Airplanes, fire suppression chemicals, strike teams, and smoke jumpers were all brought to bear on forest fires.

Federal land managers remained obsessed with containing large forest fires until about 1970. Aldo Leopold, an early environmentalist, argued as early as 1924 that wildfires were beneficial to forest ecosystems and were necessary for the natural propagation of many tree and plant species. Over the next 40 years, an increasing number of both foresters and ecologists concurred with Leopold on the ecosystem benefits of wildfires. In 1963, a group of ecologists consulted by the National Park Service released the "Leopold Report," officially titled *Wildlife Management in the National Parks,* which recommended that wildfires should be allowed to periodically burn in order to restore the environmental balance in parks. By 1968, the National Park Service had altered its fire management policies, reflecting new research. It determined that fires that were started naturally (by lightning in many cases) would be allowed to burn if they posed little risk to human life and property. The National Park Service also determined that under prescribed conditions, controlled burns would be deliberately set to both reduce the fuel available for fires and restore ecosystem balance.

By this time, research had shown that fire played an important role in the ecology of forests, which led to a turnaround in Forest Service policy—namely to let fires burn when and where this was deemed appropriate. At first this included allowing fires that were initiated by natural causes to burn in designated wilderness areas. The 1988 Yellowstone National Park fires, however, played a major role in the public's understanding of the role of fire in forest ecosystems but also the potential risk involved when fires burned completely uncontrolled. These fires were the largest in the recorded history of Yellowstone, and what started out as a number of individual smaller fires quickly spread to larger out-of-control blazes due to a combination of drought conditions, strong winds, and abundant fuel. Ultimately 793,880 acres, or 1,240 square miles, were burned over the course of several months, with the total cost of firefighting efforts reaching over $120 million. Of the initial 51 fires, 42 were caused by lightning strikes and 9 by people. While hundreds of square miles of forest were blackened and burned, not long after the fires were extinguished, tree and plant species reestablished themselves; native plant regeneration was found to be highly successful.

Since about 1990, fire suppression efforts have needed to account for increasing development in areas on the fringes of urban areas, or what is now referred to as the wildland-urban interface. On average in the western United States, the annual number of fires that burn more than 1,000 acres more than tripled between the 1970s and the 2010s, and the fire season is now 105 days longer than it was in the 1970s. The US Forest Service now spends about 12 times as much money nationally suppressing wildfires as it did in 1985. The agency's firefighting expenditure has grown to about 50 percent of its total budget, which limits the money available for land management efforts such as forest thinning and land restoration that could aid in fire suppression.

WILDLAND FIRES IN CALIFORNIA

It has become increasingly clear that wildfires in the state have become more dangerous and damaging due to a combination of the buildup of wood fuel in the forests, higher population densities at the wildland-urban interface, expansion of aboveground electrical transmission and distribution lines, and simply more people using forests for recreation. Over the past 40 years or so, the most frequent sources of ignition of wildland fires in California have been machinery that generates sparks (chain saws, brush grinders, mowers, and off-road vehicles) and overhead electrical lines, as well as arsonists and lightning.

Largest Wildland Fires

Fires in California have continued to grow in both size and intensity. The Department of Forestry and Fire Protection is the bookkeeper for wildfires in the state and has compiled records of the 20 largest wildfires, in area, through 2022 (table 7.1); the 20 deadliest in terms of human fatalities (table 7.2); and the 20 most damaging in terms of

TABLE 7.1 TWENTY LARGEST CALIFORNIA WILDLAND FIRES BY AREA

Fire Name (Cause)	Date	County	Acres	Structures	Deaths
1. AUGUST COMPLEX (*Lightning*)	August 2020	Mendocino, Trinity, Tehama, Glenn, Lake, & Shasta	1,032,648	935	1
2. DIXIE (*Power Lines*)	July 2021	Butte, Plumas, Lassen, Shasta & Tehama	963,309	1,311	1
3. MENDOCINO COMPLEX (*Human Related*)	July 2018	Colusa, Lake, Mendocino & Glenn	459,123	280	1
4. SCU LIGHTNING COMPLEX (*Lightning*)	August 2020	Stanislaus, Santa Clara, Alameda, Contra Costa, & San Joaquin	393,624	222	0
5. CREEK (*Undetermined*)	September 2020	Fresno & Madera	379,895	858	0
6. LNU LIGHTNING COMPLEX (*Lightning / Arson*)	August 2020	Napa, Solano, Sonoma, Yolo, Lake, & Colusa	363,220	1,491	6
7. NORTH COMPLEX (*Lightning*)	August 2020	Butte, Plumas, & Yuba	318,935	2,352	15
8. THOMAS (*Power Lines*)	December 2017	Ventura & Santa Barbara	281,893	1,060	2
9. CEDAR (*Human Related*)	October 2003	San Diego	273,246	2,820	15
10. RUSH (*Lightning*)	August 2012	Lassen	271,911 CA / 43,666 NV	0	0
11. RIM (*Human Related*)	August 2013	Tuolumne	257,314	112	0
12. ZACA (*Human Related*)	July 2007	Santa Barbara	240,207	1	0
13. CARR (*Human Related*)	July 2018	Shasta & Trinity	229,651	1,614	8
14. MONUMENT (*Lightning*)	July 2021	Trinity	223,124	28	0
15. CALDOR (*Human Related*)	August 2021	Alpine, Amador, & El Dorado	221,835	1,005	1
16. MATILIJA (*Undetermined*)	September 1932	Ventura	220,000	0	0
17. RIVER COMPLEX (*Lightning*)	July 2021	Siskiyou & Trinity	199,359	0	0
18. WITCH (*Power Lines*)	October 2007	San Diego	197,990	122	2
19. KLAMATH THEATER COMPLEX (*Lightning*)	June 2008	Siskiyou	192,038	1,650	2
20. MARBLE CONE (*Lightning*)	July 1977	Monterey	177,866	0	0

NOTE: Courtesy of California Department of Forestry and Fire (CAL FIRE), public domain.

TABLE 7.2 CALIFORNIA'S DEADLIEST WILDLAND FIRES BY LIVES LOST

Fire Name (Cause)	Date	County	Acres	Structures	Deaths
1. CAMP FIRE *(Power Lines)*	November 2018	Butte	153,336	18,804	85
2. GRIFFITH PARK *(Unknown)*	October 1933	Los Angeles	47	0	29
3. TUNNEL–OAKLAND HILLS *(Rekindle)*	October 1991	Alameda	1,520	2,900	25
4. TUBBS *(Electrical)*	October 2017	Napa, Sonoma, & Lake	36,807	5,643	22
5. NORTH COMPLEX *(Lightning)*	August 2020	Butte, Plumas, & Yuba	318,935	2,352	15
6. CEDAR *(Human Related)*	October 2003	San Diego	273,246	2,820	15
7. RATTLESNAKE *(Arson)*	July 1953	Glenn	1,340	0	15
8. LOOP *(Unknown)*	November 1966	Los Angeles	2,028	0	12
9. HAUSER CREEK *(Human Related)*	October 1943	San Diego	13,145	0	11
10. INAJA *(Human Related)*	November 1956	San Diego	43,904	0	11
11. IRON ALPS COMPLEX *(Lightning)*	August 2008	Trinity	105,855	10	10
12. REDWOOD VALLEY *(Power Lines)*	October 2017	Mendocino	36,523	543	9
13. HARRIS *(Unknown)*	October 2007	San Diego	90,440	548	8
14. CANYON *(Unknown)*	August 1968	Los Angeles	22,197	0	8
15. CARR *(Human Related)*	July 2018	Shasta & Trinity	229,651	1,614	8
16. LNU LIGHTNING COMPLEX *(Lightning / Arson)*	August 2020	Napa, Sonoma, Yolo, Stanislaus, & Lake	363,220	1,491	6
17. ATLAS *(Power Lines)*	October 2017	Napa & Solano	51,624	781	6
18. OLD *(Human Related)*	October 2003	San Bernardino	91,281	1,003	6
19. DECKER *(Vehicle)*	August 1959	Riverside	1,425	1	6
20. HACIENDA *(Unknown)*	September 1955	Los Angeles	1,150	0	6

NOTE: Courtesy of California Department of Forestry and Fire (CAL FIRE), public domain.

TABLE 7.3 CALIFORNIA'S MOST DESTRUCTIVE WILDLAND FIRES BY STRUCTURES LOST

Fire Name (Cause)	Date	County	Acres	Structures	Deaths
1. CAMP FIRE (*Power Lines*)	November 2018	Butte	153,336	18,804	85
2. TUBBS (*Electrical*)	October 2017	Napa, Sonoma, & Lake	36,807	5,643	22
3. TUNNEL–OAKLAND HILLS (*Rekindle*)	October 1991	Alameda	1,520	2,900	25
4. CEDAR (*Human Related*)	October 2003	San Diego	273,246	2,820	15
5. NORTH COMPLEX (*Lightning*)	August 2020	Butte, Plumas, & Yuba	318,935	2,352	15
6. VALLEY (*Electrical*)	September 2015	Lake, Napa, & Sonoma	76,067	1,958	4
7. WITCH (*Powerlines*)	October 2007	San Diego	197,990	1,650	2
8. WOOLSEY (*Electrical*)	November 2018	Ventura	96,949	1,643	3
9. CARR (*Human Related*)	July 2018	Shasta & Trinity	229,651	1,614	8
10. GLASS (*Undetermined*)	September 2020	Napa & Sonoma	67,484	1,520	0
11. LNU LIGHTNING COMPLEX (*Lightning / Arson*)	August 2020	Napa, Solano, Sonoma, Yolo, Lake, & Colusa	363,220	1,491	6
12. CZU LIGHTNING COMPLEX (*Lightning*)	August 2020	Santa Cruz & San Mateo	86,509	1,490	1
13. NUNS (*Power Lines*)	October 2017	Sonoma	44,573	1,355	3
14. DIXIE (*Power Lines*)	July 2021	Butte, Plumas, Lassen, Shasta, & Tehama	963,309	1,329	1
15. THOMAS (*Powerlines*)	December 2017	Ventura & Santa Barbara	281,893	1,060	2
16. CALDOR (*Human Related*)	August 2021	Alpine, Amador, & El Dorado	221,835	1,005	1
17. OLD (*Human Related*)	October 2003	San Bernardino	91,281	1,003	6
18. BUTTE (*Power Lines*)	September 2015	Amador & Calaveras	70,868	965	2
19. JONES (*Undetermined*)	October 1999	Shasta	26,200	954	1
20. AUGUST COMPLEX (*Lightning*)	August 2020	Mendocino, Trinity, Tehama, Glenn, Lake, & Shasta	1,032,648	935	1

NOTE: Courtesy of California Department of Forestry and Fire (CAL FIRE), public domain.

structures destroyed (table 7.3). There have been, on average, an astounding 8,292 fires annually in California in the 22-year period from 2000 to 2022. The acreage burned, however, has increased in the last decade, nearly doubling from an average of 1,063 square miles annually in 2000 to 2010, to 2,071 square miles per year in 2011 to 2021. The massive wildland fires in 2017, 2018, 2020, and 2021 significantly increased the average annual area burned for the last decade. In 2020 alone, 6,871 square miles of California burned, which is the equivalent of burning all of Santa Cruz, San Mateo, San Francisco, Marin, Contra Costa, Alameda, Napa, and Sonoma Counties. That's a big chunk of California.

To put it mildly, 2020 was an extremely bad year for wildfires in California, as it was across the entire western United States (figure 7.1). By the end of 2020, over four percent of the entire state's area had been burned, making it the worst fire season in California's recorded history. There were 74 fires of over 1,000 acres (1.56 square miles), although some of these consisted of multiple blazes that coalesced to from large complex fires. The August Complex Fire in Northern California started from 38 separate fires ignited by lightning strikes. It burned over a million acres across seven counties in the Northern California Coast Ranges (Glenn, Mendocino, Lake, Tehama, Trinity, Humboldt, and Colusa and encompassed an area larger than the entire state of Rhode Island. In terms of area scorched, this fire complex was the largest in the state's recorded history. The fire burned largely within the Mendocino National Forest, with portions spilling over into the Shasta-Trinity National Forest and Six Rivers National Forest, but also into private lands surrounding the national forests.

As has often been the case with the recent wildland fires in California, rugged terrain, combined with consistent high winds and high temperatures, made firefighting challenging, and it took almost three months to completely contain the fire. There were 935 structures that were reportedly destroyed and a single fatality, a firefighter.

The 2020 fires in California in total destroyed over 10,000 structures and had a total cost of $12.079 billion, including over $10 billion in property damage and $2.079 in firefighting costs. Not many areas escaped the devastating fires of that year, which impacted all but 5 of the state's 58 counties. The intensity of this fire season was attributed to a combination of over a century of poor forest management and higher temperatures and drier conditions from climate change. Since the 1970s, the average spring and summer temperatures in California have gone up by 1.8 degrees

FIGURE 7.1. Total aerial extent of California wildfires in 2020. *By Phoenix7777, CC BY-SA 4.0, via Wikimedia Commons.*

Fahrenheit (1 degree Celsius). In addition, the average wildfire season now lasts at least 2.5 months longer than it did in the early 1970s.

The second-largest fire in the state's history came 11 months later, in July 2021, and burned over 963,309 acres (1,505 square miles), also in Northern California in Butte, Lassen, Plumas, Shasta, and Tehama Counties. Named the Dixie Fire, it was the largest single-source wildfire in California history at the time, destroying 1,311 structures. This fire lasted for 23 days, expanding quickly over the drought-affected landscape, destroying the gold rush town of Greenville and 91 buildings. Pacific Gas and Electric subsequently disclosed that its equipment may have caused this catastrophic fire. This fire, along with many recent others, pointed out both the effects of climate change and neglected forest management, but also that the state's electric grid is highly prone to sparking wildland fires.

The third on the top 20 list is the Mendocino Complex Fire, which was a combination of two wildland fires, the River Fire and Ranch Fire. It blackened a total of 459,123 acres of Mendocino, Lake, Colusa, and Glenn Counties over three months in July 2018 before being contained on September 18, 2018. High heat, low humidity, and rugged terrain led to major challenges for fire control, with strong winds also contributing to fire growth. Hot spots persisted into the winter until final control on January 4, 2019. These fires collectively destroyed 2,526 structures and led to costs of at least $304 million, including $66 million in insured property damage and $238 million in firefighting costs (all in 2022 dollars).

The SCU (Santa Clara Unit) Lightning Complex Fire is the fourth-largest wildland fire in California history and burned through the Diablo Range, with the complex including fires in Santa Clara, Alameda, Contra Costa, San Joaquin, and Stanislaus Counties. These fires burned a total of 396,625 acres (614 square miles) from August 16 to October 1, 2020. The historic but still active Lick Observatory on Mount Hamilton was surrounded by the fire, which destroyed one building and damaged several others but did not impact the telescope domes themselves. A total of 225 structures were destroyed in this fire, with an additional 26 damaged. There were no fatalities, although six people were injured. Postfire surveys on some of the burned land indicated that the fire was mostly beneficial for the ecology of the region, despite the extent of the area scorched. This was attributed to the fire's low intensity, which spared many of the trees and rejuvenated vegetation. This assessment led CAL FIRE officials to consider increasing the use of controlled burns in Alameda and Contra Costa Counties.

Most Destructive Fires

The California Department of Forestry and Fire Protection also has kept track of the most destructive fires in the state's history based on the number of structures burned (see table 7.3). The 20 most destructive wildfires have all occurred since 1991, and only 2 of these took place before 2000. This is likely due to a combination of more development occurring in rural areas at the wildland-urban-rural contact, where encroachment has continued, and also the continuing drought and warming conditions overall, which have provided more fuel for fires. The Camp Fire in Butte County, which began on November 8, 2018, was without question the most destructive in California history, and it also ranks as the deadliest wildfire in US history since the Cloquet Fire exactly a century earlier. The Cloquet Fire was an immense blaze in northern Minnesota in October 1918 that was the worst disaster in the state's history; it left 453 fatalities, 52,000 injured or displaced people, and 38 destroyed communities. The 2018 Camp Fire was ignited by a faulty electrical transmission line, a problem that seems to have reached crisis proportions as power lines have followed development into rural, often wooded areas. Strong winds from the east drove the fire downhill into developed areas (figure 7.2) and worked against the nearly 2,300 firefighters who battled the flames.

In 2005, somewhat prescient for this disastrous fire, CAL FIRE had released a fire management plan for this region, which warned that the town of Paradise was at risk for an ember-driven conflagration similar to the Oakland Firestorm of 1991. The report stated that "the risk to the ridge communities is from an east wind-driven fire that originates above the communities and blows downhill through developed areas."[1]

FIGURE 7.2. The Camp Fire burning out of control with winds from the northeast (upper right) driving the fire towards the town of Paradise. *By Joshua Stevens, courtesy of NASA, public domain, via Wikimedia Commons.*

FIGURE 7.3. Total destruction of homes in the town of Paradise from the Camp Fire in 2018. *By Shealah Craighead, courtesy of Trump Whitehouse Archived, public domain, via Flickr.*

At its peak, the fire was burning through 2,000 acres per hour, devouring everything in its way—homes (figure 7.3), businesses, and big box stores—ultimately blackening 153,336 acres (240 square miles). Virtually the entire town of Paradise was burned, losing about 95 percent of its structures and forcing the evacuation of about 30,000 people.[2] The smaller nearby towns of Concow, Magalia, and Butte Creek Canyon were also mostly destroyed. Residents were able to grab a few possessions and their pets and then flee, driving as fast as possible on roads flanked by flames on both sides. A total of 18,804 structures were destroyed, which included homes and outbuildings as well as commercial properties.

Drought was a key factor in the spread and intensity of the Camp Fire. Paradise, which usually receives about five inches of rain by mid-November, had seen just one-eighth of an inch by that time in 2018. With the arrival of the first major rainstorm of the season, the fire was brought completely under control by November 25.

The total Camp Fire damage was estimated at over $400 billion, with insured losses estimated at only $12.5 billion and over $150 million in firefighting costs. In January 2019, Pacific Gas and Electric (PG&E), the utility company responsible for the faulty power line, filed for bankruptcy, citing liabilities from wildfires of $30 billion. In December 2019, the utility made a settlement offer of $13.5 billion for wildfire victims, which covered losses from several other fires caused by the utility in addition to the Camp Fire. In June of 2020, PG&E pleaded guilty to 84 counts of involuntary manslaughter.

The Tubbs Fire in October of 2017 ranks as the second most destructive fire in California's history, incinerating 5,636 structures and blackening 36,807 acres (58 square miles) of Napa, Sonoma, and Lake Counties. The fire started in the northern rural portion of the city of Calistoga in Napa County and was ignited "by a private electrical system adjacent to a residential structure."[3] Half of the structures lost were homes in Santa Rosa. The city's economic loss was estimated at $1.4 billion (in 2022 dollars), with five percent of the housing stock lost. At the time, this was the most destructive wildfire in the state's history, claiming the lives of 22 people in addition to the structures burned, but it was surpassed a year later by the Camp Fire. What was a surprise to many was how quickly a fire that started in a wooded area could race through an urban area and destroy entire blocks of homes almost within minutes due to weather conditions.

The Tubbs Fire was one of 21 major fires that broke out in early October 2017 in what was labeled the Northern California firestorm. These fires were also known as the North Bay Fires and the Wine Country Fires, and they scorched parts of Napa, Lake, Sonoma, Mendocino, Butte, and Solano Counties during severe weather conditions (figure 7.4). By October 14, these fires had burned over 210,000 acres (328 square miles), leading to the evacuation of 90,000 residents from their homes, the deaths of 44 people, and the hospitalization of 192 others. This wildfire event was one of the deadliest in the United States over the past century.

The 1991 Tunnel Fire in the Oakland hills, also named the Oakland Firestorm of 1991 and the East Bay Hills Fire, was the third most destructive in California history. This blaze, like the fires described above, also occurred at the wildland-urban interface. It burned through the hills of northern Oakland and southeastern Berkeley over the weekend of October 19–20, 1991. While the total acreage burned was not great compared to the Camp and Tubbs Fires (1,600 acres), destruction was extensive because of the high density of development in these hills. The firestorm killed 25 people, including a fire battalion chief and a police officer, and injured 150 others. Losses included 2,843 single-family dwellings and 437 apartment and condominium units. The economic cost was estimated at $1.5 billion ($2.8 billion in 2022 dollars).

FIGURE 7.4. October 8, 2017, satellite image showing smoke from the North Bay and Wine Country Fires. © *The European Space Agency (ESA), CC BY-SA 3.0 IGO.*

The fire started on Saturday afternoon, October 19, from an incompletely extinguished grass fire in the Berkeley Hills, northeast of the intersection of Highways 24 and 13, a half mile north of the west portal of the Caldecott Tunnel. The initially small grass fire (5 acres) was thought to be extinguished by Saturday evening. It restarted as a brush fire on Sunday morning at 11:00 a.m. on a steep slope and spread quickly southwest due to wind gusts of up to 65 miles per hour. Local and regional firefighters were soon overwhelmed by the erratic and extreme fire behavior due to the strong winds. Embers were blown by the high winds, starting new spot fires, and within less than an hour the fire had crossed both Highway 24, an eight-lane freeway, and all four lanes of Highway 13 (figure 7.5). This quickly ignited hundreds of homes in the Forest Park neighborhood on the northwest edge of the Montclair district and in the upper Rockridge neighborhood (figure 7.6). The fire just reached to the edge of Piedmont, where it burned some municipal property, but homes were spared. The winds were so

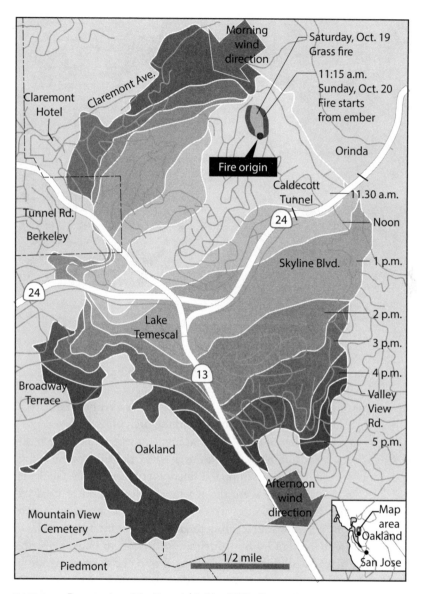

FIGURE 7.5. Progression of the Tunnel / Oakland Hills Fire on October 19–20, 1991.

FIGURE 7.6. Homes destroyed in the Tunnel / Oakland Hills Fire. *By Robert Eplett, courtesy of the California Governor's Office of Emergency Services, public domain, via Flickr.*

strong that they blew debris across the bay into San Francisco and onto the field at Candlestick Park, where the San Francisco 49ers were playing the Detroit Lions. At 9:00 p.m., the fierce winds abruptly came to a halt, giving firefighters a chance to contain the fire.

There were a number of problems that immediately became apparent as firefighters from different agencies arrived at the fire. Outside fire teams discovered various equipment incompatibility issues, such as hydrants with outlets that were the wrong size for the hoses used by neighboring counties. Oakland was also unable to communicate with a number of mutual aid resources, due to antiquated equipment and lack of access to statewide radio frequencies, resulting from budget restrictions in prior years. Adding to the challenges were the narrow, winding roads through the area that were lined with parked cars, including many in front of fire hydrants, that made access difficult for fire trucks. The flames also destroyed power lines to 17 different pumping stations in the Oakland water system, which prevented the refilling of empty reservoirs, causing the firefighters to literally run out of water. In retrospect, all of these problems could have been avoided.

Homes that burned in the firestorm along the slopes of the Oakland hills have since been rebuilt, but weather conditions and the vegetation aren't much different now, except that with an extended drought in California, the temperatures are getting warmer and the landscape is now drier with lots of dead trees. Concern has grown in the subsequent 32 years about a repeat fire in the same area. All new or significantly remodeled houses in California today are required to have interior fire sprinklers. While that's a good precaution to limit the spread of interior fires, it would seem to make a lot more sense in these wildland-urban transition areas to install external roof sprinklers.

Close behind the Tunnel Fire, or Oakland Firestorm, in its destruction was the Cedar Fire that burned over 273,246 acres (427 square miles) of San Diego County in October of 2003. This fire destroyed 2,820 structures, including 2,232 homes, and led to 15 fatalities. The fire started in the Cuyamaca Mountains in the Cleveland National Forest on the early evening of October 25. Following the fire, investigators determined that a novice hunter who was lost had set a fire to signal possible rescuers. Unfortunately, the pile of brush he lit on fire quickly spread because of the heat, low humidity, and low moisture content of the surrounding vegetation.

Initial response was immediate, with 10 fire engines, two water tenders, and two hand crews sent within 10 minutes of the first report. In another 20 minutes, 320 additional firefighters were en route. Then the first system failure occurred. A sheriff's helicopter equipped with a large water bucket was dispatched to drop water on the fire and was only minutes away from the blaze when a Forest Service fire chief cancelled the water drop because policy required halting aerial drops 30 minutes before sunset. This decision was later severely criticized by the public, media, and elected officials. This action and other controversies led to investigations lasting several years that came to some striking conclusions. Although the fire conditions and severity should have been expected, the responsible agencies weren't properly prepared when the fire started, and radio communications exacerbated the problem. A report issued by an investigating committee stated, "Disorganization, inconsistent or outdated agency policies that grounded aircraft or caused other problems, and planning or logistics in disarray also marked the preliminary stages of the difficult, dangerous firefighting."[4]

The rapid spread of the wildland fire (3,600 acres per hour) was driven by Santa Ana winds blowing out of the mountains towards the coast. At the time the fire started, there were already at least 11 other wildfires burning in Southern California. Between 5:37 p.m., when the fire was reported, and midnight, over 5,000 acres had already been

FIGURE 7.7. The Cedar Fire of October 2003. *By John Gibbons © San Diego Union-Tribune via ZUMA.*

consumed. Fifteen people, including one firefighter, died in the fire. Many of those killed were trapped by the flames driven by the 60-mile-per-hour winds that drove the flames faster than residents could flee. At least 10 individuals were trapped in their vehicles while trying to escape the flames (figure 7.7). As of January 2022, the Cedar Fire remained the ninth-largest wildfire, the sixth deadliest, and the fourth most destructive in California history, causing just over $1.3 billion in damages.

In 2000 there were 120 days with high risk for wildfires in California. By 2050, the state is projected to have 24 more days each year with high risk for wildfires. This is not just a California problem, however; the number of homes in the United States at risk of wildland fires has gone up dramatically in recent years, from 30.8 million in 1990 to 43.4 million in 2010.

One of the lessons we have learned from the history of wildland fires in the state is that certain areas are particularly prone to repeated fires. The Oakland hills, for example, have suffered through fires in 1923, 1931, 1933, 1937, 1946, 1955, 1960, 1961, 1968, 1970, 1980, 1990, 1991, 1995, 2002, and 2008. That's 16 fires in 100 years, or one every 5 years on average. Other areas particularly prone to wildfires include parts of Orange, Riverside, San Bernardino, and Los Angeles Counties. Orange and San Bernardino Counties share a border that runs north to south through the Chino Hills State Park, where the landscape ranges from coastal sage scrub, grassland, and woodland to areas of brown and sparsely dense vegetation made drier by droughts or hot summers. The valley's grass and barren land can easily become susceptible to dry periods and drought, thereby making it a prime spot for brush fires and conflagrations, many of which have occurred since 1914. The hillsides and canyons there have seen brush fires or wildfires in 1914 and in the 1920s, 1930s, 1940s, 1950s, 1960s, 1970s, 1980s, 1990s, 2000s, up to the present. These four counties are home to 17.6 million people, many of whom use the national forests, state parks, and other open spaces for various types of recreation. Whether hiking, camping, or in other outdoor pursuits, this many people in these large outdoor spaces, with their typically dry summer and fall conditions, simply increase the likelihood of an accidental fire.

SOME FINAL THOUGHTS ON WILDLAND FIRES IN CALIFORNIA

California is unfortunately home to 5 of the 15 largest fires in US history, due to a combination of factors: vast expanses of wildlands in close proximity to large urban areas; an expanding population and development that has progressively encroached into these wooded or chaparral-covered hillsides and canyons; persistent drought conditions along with rising temperatures in recent decades that have desiccated the vegetation, leaving it more susceptible to combustion; frequent lightning storms that can ignite dry vegetation; steep terrain with limited access, making firefighting more difficult; and the understandable desire of many homeowners living at the wildland-urban interface to be surrounded by trees and other vegetation, which provide some privacy but increase the fire hazard and allow for the rapid spread of fires once they start. Adding to these conditions are the challenges that have affected some of the firefighting agencies themselves, as discussed earlier in this chapter: outdated communication systems, incompatible firefighting equipment, and lack of preparation for the inevitable fires.

Eighteen of California's 20 largest wildland fires have occurred since 2000, and 12 of these have occurred in the past five years. Eleven of the 20 deadliest fires (in terms of loss of life) have also occurred in this same time period. Eighteen of the 20 most destructive fires (in terms of structures lost) have also taken place since 2000. With a planet continuing to warm, and California having a Mediterranean climate, we can expect to see temperatures continue to rise, more frequent and longer droughts, and also population and development expanding at the wildland-urban interface. These all increase the risks of future large and destructive wildland fires.

The increasing use of controlled burns in some jurisdictions has helped to eliminate the accumulated fuel in some areas, thereby reducing the risks of large fires. A combination of restrictions on either rebuilding or new construction in indefensible areas, rooftop fire sprinklers, and rising fire insurance premiums or policy cancellation in fire-prone areas will also contribute to reducing future risks. In 2023, two of the major home insurers decided not to insure homes in California, due to increased risks and losses in recent years.

8

Landslides, Rockfalls, and Debris Flows

INTRODUCTION

Living in the Santa Cruz Mountains at the time of a 100-year, 24-hour rainstorm in early January 1982, I was aware that the rainfall had been extreme by any measure. I did notice that morning that my rain gauge had overflowed but initially thought that perhaps I hadn't emptied it after the previous storm. Listening to the local radio station that morning of January 4, I heard a report that an entire mountainside had collapsed above Ben Lomond in the San Lorenzo Valley and that there were major concerns with homes and people having been buried. I made my way through roads, partially blocked with mud, rocks, and downed trees, to Love Creek; and when the road was completely closed, I got out and hiked the last quarter mile to the base of this massive slide. I was stunned by the sheer magnitude of the pile of mud, rock, and trees, which had dammed the creek, where a lake was rapidly forming (figure 8.1). There were remnants of homes, furniture, clothes, and Christmas ornaments mixed up with the mud and trees that had come to rest at the bottom of the mountain.

Friends and families were gathered where two backhoes were digging away at the massive pile of 600,000 cubic yards of mud and debris, searching for any possible survivors. Ten individuals lost their lives that night, four were never recovered. One woman survived while her roommates died, making the Love Creek slide one of the most devastating in California's history. Hiking up the muddy slide mass that morning,

FIGURE 8.1. The January 1982 Love Creek slide in the Santa Cruz Mountains involved about 600,000 cubic yards of mud and rock. © *1982 Gary Griggs.*

which was still unstable, I was trying to both understand what had happened and also be aware of the continuing risk of additional failures on the hillside, which became an area of controversy for months to follow. Additional headscarp cracks were soon discovered and mapped adjacent to the failed area, leading to the condemnation of a number of additional homes as being unsafe to occupy. While the trauma and fatalities from that tragic event have slowly begun to heal, nearly 40 years later, the Love Creek slide scar is still there on the hillside.

With increasing frequency, the sounds of snapping two-by-fours, shattering glass, and the cries of the suffering homeowner, and ultimately the arguments of the lawyers in court, have been heard as an accompaniment to an old and common geologic process—the landslide. Not long ago, an entire hillside could slide into a canyon, or a slab of sea cliff would collapse into the waves, and no one would much care. Now, however, as the cliffs along the California coast and the rural and suburban hillsides have become increasingly popular with housebuilders and homeowners, slope failures of this sort have become frightening, devastating, and expensive events. While rockfalls, landslides, slumps, and debris flows don't garner the press and attention of earthquakes and floods, they are surprisingly common and widespread.

FIGURE 8.2. Repair work at the massive Mud Creek landslide along Highway One in Big Sur, July 2017. *Courtesy of US Geological Survey, public domain.*

Along Central California's rugged Big Sur coast, massive landslides and extended road closures have become a way of life in winters with heavy rainfall for those living, working, or traveling along this iconic coast (figure 8.2). In the Sierra Nevada, rockfalls tend to be the most frequent type of slope failure, and in the subdivided and urbanized hillsides surrounding the greater Los Angeles and San Francisco Bay metropolitan areas, particularly after major wildland fires, it is debris or mudflows that get our attention and do the most damage.

Landslides are just one of a number of types of downslope movements or slope failures, consisting of soil, rock, and other materials driven downhill by gravity, that are often initiated or accelerated by excess rainfall or water or, less frequently, earthquakes (figure 8.3). Some slides may involve only a truckload of wet soil or rock, while in other cases, deep failure may involve entire hillsides, hundreds of thousands of cubic yards of earth and rock, and dozens of homes. Some types of slope failures are very slow, moving only inches per day, while others, such as debris flows or mudflows, move faster than any of us can run. The threats and risks vary, then, as to the amount and type of material moving towards your home or car or you. None of these events is going to be beneficial, but some events may be far more destructive and dangerous than others.

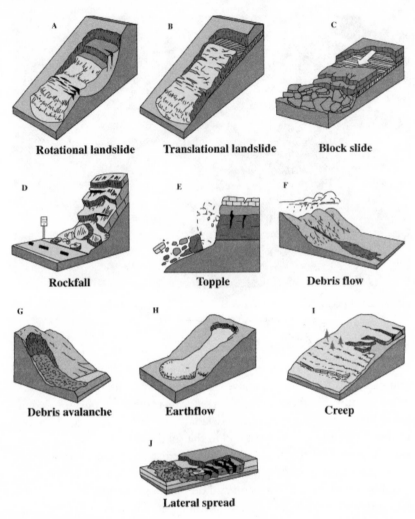

A · Rotational landslide   B · Translational landslide   C · Block slide

D · Rockfall   E · Topple   F · Debris flow

G · Debris avalanche   H · Earthflow   I · Creep

J · Lateral spread

FIGURE 8.3. Different types of mass downslope movements. *Courtesy of US Geological Survey, public domain.*

While not always obvious in many of the state's more mountainous regions because of the forest or brush cover, landslides and other types of downslope movements are a widespread phenomenon affecting the shape of California's youthful landscape (figure 8.4). For many, it may only be when we see cracks offsetting the asphalt, or rocks or mud falling or oozing onto the roadway and blocking traffic, that we even become aware of this all-too-common process. Because of the high

FIGURE 8.4. A large, but somewhat subtle, earthflow in the hills of Contra Costa County. Mass movements can play a major role in landscape evolution in California but are not always evident. © *1982 Gary Griggs.*

rainfall in some of the state's mountainous areas, 50 or 60 inches in most years, the scars left behind by slope failures are quickly vegetated and within a few years are usually unrecognizable. Even to the trained geologist, evidence of previous landslides or mudflows is not always obvious or apparent, making it challenging to assess the stability of a hillside and the potential for future failure.

The more years that have passed since a hillside has slid or slipped, the more the landforms and vegetation conspire to heal the land and cover the evidence. This may be beneficial to those trying to sell mountainside property, but in the long run, information of this sort on areas prone to failure is important to consider before buying or building in steep terrain. Having a modest-size, 30-cubic-yard debris flow (roughly three loads from a dump truck) collide with a typical wood-frame dwelling is like driving that same truck into it at 15–20 miles per hour. Most homes don't survive impacts of this sort very well, and it's advisable to do everything we can to avoid this happening. Unfortunately, as with most geologic hazards or natural disasters, events like this occur irregularly and somewhat infrequently, and the memory of these usually fades fairly quickly. We tend to share a collective amnesia for natural disasters and try to get over them and move on as quickly as possible.

If you are considering a move and buying property or a home, whether in the hills, mountains, or directly on the coast, there are several different types of hillside failures that you should consider before investing your life savings.

ROCKFALLS

Rockfalls are perhaps the simplest type of slope failure to identify and observe and also the easiest to understand. These events occur when loose or weak rock on a sea cliff, or along a steep road cut or canyon wall, breaks loose and falls, rolls, or slides downhill. This process is very common along the high and near-vertical sea cliffs of northern Monterey Bay in Santa Cruz County (figure 8.5). Much of the rock along these cliffs is relatively weak sandstone and mudstone, and due either to undercutting or impact by waves, excess rainfall, or shaking during an earthquake, the rock breaks loose along weakness planes and falls to the beach below.[1] This can be a serious hazard if you happen to be either walking or sitting on the beach beneath these cliffs. Even a casual visitor would probably recognize the large rocks at the base of the cliff and hopefully have the common sense not to spread out their towel among the rocks, but this is not always the case. A greater concern exists for those who own homes on the cliff top and are gradually losing their backyards and patios with each chunk of rock that breaks loose.

On October 17, 1989, we learned that large earthquakes can also trigger failures of sea cliffs. Seismic shaking during the 6.9 magnitude Loma Prieta earthquake produced bluff failure from San Francisco to central Monterey Bay (figure 8.6). One of the seven deaths in Santa Cruz County during that earthquake occurred at a beach along Santa Cruz County's north coast when a sunbather was buried under a rockfall from the overhanging cliff. The bluffs throughout Capitola and Rio Del Mar also experienced widespread collapse due to the severe shaking. Cracks appeared at the bluff top beneath foundations and patios, large masses of rock and soil broke loose and slid or fell to the beach below, and some of these blocked access to and damaged homes and buried cars at the base of the bluff in places like Pot Belly Beach and Las Olas Drive just south of New Brighton State Beach.[2]

Six apartment units on Depot Hill in Capitola had to be demolished, as the concrete caisson support system was undermined by cliff failure, which led to cracking of the foundations and walls. Three other bluff-

FIGURE 8.5. Rockfall from the sea cliff at Capitola, northern Monterey Bay, which presents a hazard to beachgoers as well as cliff-top homeowners. © *1970 Gary Griggs.*

FIGURE 8.6. Massive bluff failure in Daly City from the 1989 Loma Prieta earthquake. The scar of an older landslide can be seen to the right. © 1989 Gary Griggs.

top homes were also ultimately demolished as their ocean view sites were cracked beyond repair by the shaking accompanying the earthquake. It's important to recall that the epicenter of that shock was only a few miles inland from these sites and that shaking was intense.

Highway One at Waddell Bluffs, just south of Point Año Nuevo and the San Mateo County line, is another site where the weak rock in the steep, 300-foot-high bluffs continues to fail and then roll and slide downhill towards the highway. The California Department of Transportation (Caltrans) built a ditch alongside the road to catch the rolling and falling rocks; sometimes this works, but historically if the ditch hadn't been emptied for a while, large rocks frequently would bound out onto Highway One. Despite the warning signs at either end of this hazardous stretch of highway asking you to "Watch for Rocks," there is no follow-up message advising you what to do if you see a rock racing downhill towards your car. Vehicles have been damaged and waylaid by colliding with rocks that either have come to rest on the roadway or are still moving across the road. A man lost his life at Waddell Bluffs in 1982 when a 200-pound rock rolled downslope and through the trench and became airborne before breaking through the front window of a truck. Following a lawsuit against Caltrans, the maintenance of the trench and its safety have improved, with removal of the rocks and debris more regularly and also the construction of a heavy cable barrier to prevent rocks from entering the highway.

The mountain passes cutting through the Sierra Nevada are sites of frequent rockfalls due to the often steep-sided road cuts, the seasonal freezing that can wedge large rock masses loose, and the extreme seasonal temperature ranges that can lead to expansion and contraction of rock masses, thereby decreasing the stability of the slope. Although the Sierras have a limited history of large landslides, because of the general overall stability of the granitic rocks, there have been prehistoric rockfalls and rockslides that could be very damaging if they occurred today in the more densely populated or traveled parts of the area. Several documented very large rockfalls and rockslides were initiated by either long periods of unusually wet weather or strong seismic shaking, although in some cases there were no obvious triggering events. Large rockfalls have frequently blocked and closed the major mountain highways crossing the Sierra Nevada until the rocks have been moved by heavy equipment or blasted into movable blocks.

LANDSLIDES AND SLUMPS

Landslides and slumps may include rock, soil, or a mixture of the two, as well as anything on the overlying landscape, including trees, houses, roads, or cars. They can be small or very large and are distinguished by usually moving relatively slowly. These events involve a mass of material that fails along a plane or surface and can move as a relatively intact mass but can also break apart as it moves downhill. If the failure surface is smooth or planar and the material is relatively undisturbed, it is called a slide because the material glides slowly or slides downhill (see figure 8.3). More commonly, the failed hillside material breaks up into a series of separate blocks and is distorted and rotates as it slips downhill and is called a slump. Slides and slumps can be a few feet to hundreds of feet thick and may involve dozens of acres of hillside, leaving behind deep cracks and a jumbled mass of soil and rock. Although geologists have divided downslope movement of rock and/or soil into distinct categories, in many cases these may be transitional or the event may include several types of movement. Some slumps, for example, may transition to flows of mud and rock at the base of a slope.

Hillsides are more prone to failure if there is excess water involved. Water in the pores of the soil or rock weakens the material by reducing its internal friction, so most landslides or slumps occur during the winter months after prolonged rainfall, which can penetrate deeply into the subsurface. Deep-seated slope failures are usually initiated by sustained rainfall over weeks or months that percolates deep enough to destabilize a hillside to the point of failure.

After a landslide or slump has taken place, often during a very wet winter, the resulting landforms are easily recognized. A scarp or low cliff and curved or concentric cracks typically mark the upper portion or head of the slide (see figure 8.3). The main portion of a landslide is often distorted and broken up with cracks along the edges or margins, and the lower end, or toe, is often seen as a bulge at the base of the hillside. A rotational slump, on the other hand, may have a surface that can be relatively flat, although any trees present have probably been tilted, often uphill, sometimes downhill, but probably no longer standing vertical.

These flattened areas along the middle sections of these large slumps on otherwise steep hillsides may look like perfect building sites to the innocent buyer or builder and, as a result, are frequently developed. On the positive side, an area that has already failed and readjusted to the forces of gravity is typically more stable than it was before. But unfortunately, adding the load of a house and putting excess water into the ground from a septic tank and leach field, or from rain gutter or driveway runoff, may lead to renewed instability and downhill movement.

MUD AND DEBRIS FLOWS

Debris flow and *mudflow* are terms often used interchangeably and describe slope failures that are fluid rather than solid and therefore flow downhill, in contrast to the slower-moving landslides or slumps just described. While these flows can be muddy, they usually also carry very coarse material such as cobbles and boulders the size of automobiles. These fluid flows often begin within a depression or swale on a hillside where surface and/or subsurface runoff has accumulated and weakened the soils and weathered bedrock. Once these surface materials have become saturated with water, they can liquefy and flow downhill along a fairly narrow corridor or path, perhaps only 10 to 25 feet wide, and then spread out in a fan shape when they reach the base of the hillside or a flatter area. While mudflows are common in areas where the vegetation has been removed through forest or brush fires, or through logging, they also occur on undisturbed hillsides as normal events or processes under sustained rainfall.

Debris flows or mudflows form when several conditions are met. First, there must be an accumulation of at least several feet, or often more, of permeable soil and weathered bedrock on top of more impermeable bedrock, and all of this on a relatively steep slope. Second, failure of this material takes place when there has been a long period of saturating precipitation followed by high-intensity rainfall, which saturates the soils and weathered material resting on the bedrock and converts it to a fluid or mud. The impermeable bedrock at depth stops the rainfall from penetrating farther. This causes the water table to rise in the soils and changes its condition to that of mud or a viscous fluid. When this happens, the soils are no longer stable on moderate to steep slopes, and they start to ooze and then flow downslope. If the slope is steep enough and the material is fluid enough, the mudflow begins to approach the consistency of floodwater and can move quite rapidly. With a flatter slope and less water, the material will flow more slowly and probably not create as much damage.

Some mudflows may contain only 20–30 cubic yards of material, just a few dump truck loads, and take place on an unpopulated hillside in the middle of nowhere. While these are interesting events to geologists, they may have no significant impact other than modifying the hillside. At the other extreme, mud or debris flows may contain hundreds or thousands of cubic yards of material, take place within or above populated areas, and be responsible for both property destruction and also loss of life. These flows are potentially more dangerous than other types of mass downslope movements because they may move at speeds of up to 30 miles per hour—faster than people can run. They can cause considerable damage when they impact houses or other structures, because of the velocity of the mud and debris and therefore its destructive momentum (figure 8.7).

FIGURE 8.7. This debris flow came down from steep, recently burned slopes, buried most of the village of Big Sur, and picked up and destroyed vehicles in 1972. © *1989 Gary Griggs.*

HISTORY OF LANDSLIDES IN CALIFORNIA

Slope Failures from the Great April 18, 1906, San Francisco Earthquake

The severe shaking produced by the great 7.9 magnitude 1906 San Francisco earthquake induced the earliest well-documented major landslide events in the greater San Francisco Bay area. The April 18 quake followed a winter of significantly above-average rainfall, which left the soils and hillsides weak and susceptible to failure.

What has often been overlooked in the accounts of this earthquake is the flooding and landslides initiated by the very heavy rainfall that preceded the April 18, 1906, event. Boulder Creek, high in the San Lorenzo Valley, which typically has the highest rainfall totals in the Santa Cruz Mountains, had been deluged by 55.7 inches of rain during January, February, and March, along with 16 inches in the four previous months. In the week between January 11 and 28, 1906, Wrights (just east of Highway 17 near Summit Road) was hit with 27 inches of rain. Landslides, rocks, and mud blocked sections of the railroads and roadways at many locations throughout the Santa Cruz Mountains. Bridges were washed out from flooding in Soquel. One account recalled, "The Loma

Prieta Lumber Co.'s Mill at Hinckley Creek, 7 miles above Soquel, was swept away by the storm Thursday night (Jan. 18) and hardly a trace of machinery, dwelling houses, barns, skid roads or wagon roads remain. Eleven men, including J. W. Walker, the foreman of the mill, camped all night Thursday on top of the mountain for safety and fortunately no one was injured by the wash out. Some of the debris was found on the beach near Capitola."[3]

Three months later when the big earthquake occurred, very large landslides and debris flows that moved entire hillsides and forests downslope were common throughout the Santa Cruz Mountains. Although population densities in the hills were far lower than today, the locations of the particular hillsides that failed were the problem. The slopes were still very wet and the Loma Prieta Mill on Hinckley Creek, which had been seriously damaged in the January rainstorms, was struck with a devastating landslide. Nine men died as they slept in their bunkhouses when the mill was buried by a wet mass of earth 100 feet deep, while several hundred feet away others were spared. The speed of the landslide was described as "extraordinary"; it completely dammed the creek and created a lake 50 to 60 feet deep.

On Deer Creek (a tributary of Bear Creek that drains the west slope of Castle Rock Ridge in the Santa Cruz Mountains), a large landslide started from near Grizzly Rock and slid westward but changed its direction 60 degrees or more farther toward the creek. The shingle mill and houses in the creek bottom below the slide were buried under a reported 50 to 100 feet of earth, and two people were killed.

In addition to the slope failures in the mountains, the sea cliffs also responded to the severe shaking. In Capitola, it was reported that "much earth fell from the bluffs near the town, but there was no appreciable effect on the surf . . . a continuous cloud of dust rose along the cliffs between Castro's Landing [now called Rio Del Mar] and Santa Cruz."[4]

Landslides and earthflows resulting from severe shaking of the saturated ground were also widespread in the coastal hills of San Mateo and Marin Counties. Fortunately populations in these rural areas were very low at that time, so there were no injuries or fatalities, and damage was also minimal.

1933–34 Crescenta Valley Mudflow/Flood, Los Angeles County

Beginning on New Year's Eve 1933 and extending into New Year's Day 1934, the communities of La Crescenta–Montrose, La Cañada, and

Tujunga in the Crescenta Valley, north of Los Angeles, were inundated by water and mud. Wildfires had swept through the San Gabriel Mountains above the communities in November 1933, burning off most of the chapparal vegetation. Late fall fires are all too common in California's typically dry coastal mountains and often set the stage for winter mud and debris flows. Vegetation is effective in holding soil in place and reducing the impact of intense or prolonged rainfall, so when it has burned, the probability of mud or debris flows increases dramatically.

The last week of December 1933 brought a series of storms that dropped 12 inches of rain over the burned-over hillsides. These storms were followed by more heavy rain on New Year's Eve. There was concern, following the fires, with potential flooding and debris flows coming out of the mountains, which led the Civilian Conservation Corps to build earthen dams in the drainages to capture or reduce flows to the valley below. These did not do the job, however, and about midnight on December 31, the water and debris from the weakened slopes collapsed the newly built earthen dams and sent millions of tons of mud and debris into the neighborhoods below, taking the lives of about 40 people.

A number of Montrose residents took shelter at the American Legion hall, but it was in the path of the flows and was destroyed, taking 12 lives. More than 400 homes were destroyed in La Cañada, La Crescenta, Montrose, and Tujunga by the mud and debris flows, leaving hundreds without homes. Parts of Foothill Boulevard were buried under 12 feet of mud, boulders, and debris, which completely covered cars. Following this event, the Los Angeles County Department of Public Works and the Army Corps of Engineers constructed a system of catch basins and storm drains designed to avoid a repeat of the 1933–34 disaster.

1956 Portuguese Bend Landslide, Palos Verdes Peninsula, Los Angeles County

The Portuguese Bend landslide along the Southern California coast of the Palos Verdes Peninsula is an active, slow-moving mass of rock and soil that extends from the coastal hills to the shoreline (figure 8.8). This slide may be the largest and most active landslide in California. The name for this section of coast came from the original Portuguese whalers who used this area for a whaling station, and reportedly for a smuggler's hideout. The underlying sedimentary rock geology is inherently prone to failure, and sliding has taken place over thousands of years.

FIGURE 8.8. The massive Portuguese Bend landslide in 2006. © *2006 Bruce Perry.*

The landslide is massive, encompassing about two square miles (1,280 acres), and has moved along a relatively smooth basal failure surface at an average depth of about 100 feet. While unstable conditions have existed for many hundreds of years, the modern history of movement began in 1956, coincident with the construction of a road along the crest or top of the ancient landslide complex. During road construction, about 235,000 tons of material was excavated and dumped on the hillside, along with hundreds of thousands of gallons of water. Seepage of the water into the subsurface essentially lubricated a layer of bentonite clay, which was sloped or dipped downhill towards the shoreline at about six degrees.

Another probable factor in the activation of this ancient slide was the construction of 170 homes on and above the landslide in the early 1950s, despite the knowledge that the area was the site of an older landslide. This was not an uncommon practice in the hillsides surrounding Los Angeles during the era of rapid growth in the 1950s and 1960s, when grading codes were essentially nonexistent.

The area wasn't served by a sewage system, so each home had to have its own on-site cesspool or septic tank system. This introduced a significant volume of water into the subsurface year-round. Homeowners also planted lawns and gardens, which required considerable water in the semiarid Southern California climate. This percolated into the subsurface along with the fluid from the septic systems and served to lubricate the landslide plane. A rise in the water table had been documented over the period from 1957 to 1968 beneath part of the landslide. Another factor in the ongoing movement is the continuing wave erosion at the base of the slide on the shoreline, which leads to instability and renewed seaward movement.

On August 17, 1956, the initial signs of slope failure were first noticed in the cracking of the foundation of a recently built home. The cracks were patched, but they reopened a few days later. A major road running through the area then began to crack, and by September the crack had grown in width to four inches, and it continued to expand through mid-September. A monitoring system was then established, which recorded three to four inches of displacement per day. This slip had decreased to about three-quarters inch per day by the end of October 1956. Movement continued, with enlargement of the slide mass by failure of adjacent blocks of earth, and by September 1969, an estimated 60 million tons of material were moving downslope towards the shoreline. Between 1962 and 1972 the velocity varied slightly but averaged about 0.4 inch per day. The slide mass has continued to move. Survey markers on the eastern side of the slide moved 237 feet horizontally in the 20 years between October 1956 and October 1976, and 382 feet horizontally between October 1976 and October 1986. This converts to about 0.8–1.25 inches per day for this latter 10-year period.

There were originally 170 homes on the landslide, but 100 were destroyed by continued slide movement, and only 30 remain today. The Portuguese Bend clubhouse, restaurant, and pool were also destroyed. Roads have been rerouted, and the main road crossing the slide, Palos Verdes Drive, is constantly under repair. Movement continues, although at a much slower rate, about an inch per day. Some homes continue to be occupied, although regular leveling with hydraulic jacks and repair and maintenance of utilities is also required for them to remain habitable.

March 1958 Pacific Palisades Slide, Santa Monica

The steep coastal bluffs in the Pacific Palisades area of Santa Monica have a long history of failure extending back at least to the late 1800s. The city of Santa Monica was established on the bluffs, or palisades, overlooking the ocean in 1875. A stagecoach road was built at that time along the base of the sea cliffs to connect Santa Monica with the ranches and homesteads to the west along what was to become the Malibu coast. Accounts by the early users of the road reported landslides occurring during the rainy seasons of 1884 and 1889. Local residents recall more than 30 landslides having closed the Pacific Coast Highway in the 100 years from 1911 to 2012. At least eight homes were destroyed during this period, and one fatality was reported in a massive 1958 slide, which was subsequently referred to as the killer slide.

On March 27, 1958, during a heavy rainstorm, a high school student was driving his dad to work in a very small Isetta—a microcar, even smaller than a Smart car—when a huge mass of earth and mud, later estimated at about 160,000 cubic yards, came down from the steep bluff above the Pacific Coast Highway. The little car was so light that it was buoyed up by the earthflow, which carried them along like a piece of driftwood, and they ended up getting dumped on the beach, miraculously uninjured.

Four days later, a second large slide took place below Via de las Olas, where Vaughn Sheff, the district highway superintendent, was directing cleanup work on the earlier slide. He was ready to reopen the highway when an estimated 480,000 cubic yards (about 48,000 dump truck loads) of rain-loosened earth and rock broke loose from the hillside above and slid quickly down the bluff below Via de la Paz, covering the highway and burying Sheff with over a hundred feet of material. While there were at least two dozen workmen on the site, only Sheff was caught in the slide. His body was recovered seven hours later. While the 1958 slide was one of the most widely reported on because of the fatality, slides have continued to occur along these steep bluffs. A large landslide at the Palisades in June 1965 destroyed three homes, one apartment building, and part of another.

November 1972 Big Sur

The Molera Fire swept through the Big Sur area during the late summer of 1972, burning the vegetation over 4,000 acres (6.25 square miles) of steep coastal mountains. These slopes usually receive about 40 to 60 inches of rain annually. Short periods of very intense rainfall followed long saturating rain during the following November and led to extensive mud and debris flows on the steep burned-over slopes. The debris flows, estimated at 10,000 cubic yards (about 1,000 dump truck loads), moved rapidly down the steep slopes along the narrow drainages. The rocks and mud covered Highway One and partially buried the pictur-esque village of Big Sur, including houses, businesses, automobiles, and mobile homes (see figure 8.7). Boulders up to 10 feet in diameter and redwood trees up to 3 feet across were carried along by the flows, along with the mud, which hardened to the consistency of concrete, and had major impacts on the structures and vehicles in the path of the flows. Although businesses were impacted for months and Highway One was closed for days, there were no injuries or fatalities.

1978 Bluebird Canyon, Laguna Beach

Early on the Monday morning of October 2, 1978, a landslide began slowly moving downhill in the Bluebird Canyon area of Laguna Beach, an area of historic slope failures. The landslide was a large block glide that covered a 3.5-acre area and remained relatively intact as it moved down the slope. It carried 19 homes and parts of 14 others down the side of Bluebird Canyon at an initial rate of about 40 feet per hour, which slowed gradually to a few inches per day. Ultimately, it was deter-mined that 50 homes had been destroyed or affected. It also forced the evacuation of 300 more homes and completely destroyed a major street.

Geologists who studied the failure believed that the landslide was actually part of a larger, much older slide, which is not an uncommon situation. The bedrock conditions in the subsurface indicated that the bedding or layering in the sedimentary rock was dipping or sloping downhill, which is a typical recipe for instability and failure. Ten days after the initial movement, on October 12 an additional section of the slope gave way, taking another house with it. In addition to the unfavor-able bedding slope, there had been unusually heavy rains during the previous winter and spring, and the water had taken months to perco-late downward to the bedding plane where failure finally occurred.

Following a detailed geotechnical investigation, a major slope stabi-lization effort was undertaken in 1978–79, consisting of regrading the entire hillside with 340,000 cubic yards of soil reconfigured. A new storm drain system was installed—a horseshoe-shaped buttress of com-pacted fill and 66 steel "soldier piles" driven into the margins of the buttress and tied together with welded steel beams. Twenty-seven years later at 7 a.m. on June 1, 2005, an area adjacent to the 1978 slide failed, and another 22 homes were destroyed, 11 more were damaged, and an additional roughly 350 homes were evacuated along with over 1,000

FIGURE 8.9. Damage from the 2005 Bluebird Canyon landslide. *By Pam Irvine courtesy of California Geological Survey, public domain.*

residents (figure 8.9). Homes, garages, cars, and streets were carried downslope. This failure was initiated by elevated groundwater levels from the high rainfall of the prior winter. This is not uncommon for deep-seated landslides, where it may take months for the excess winter rainfall to percolate down to a potentially unstable layer in the subsurface. In response, a $33 million, 2.5-year hillside stabilization project was carried out at the base of the slope. Following this most recent slide, a communications specialist for the Insurance Information Network of California stated that landslide damage isn't really an insurance issue, because the standard homeowner's policy doesn't include land movement, and that landslide insurance is almost nonexistent in the state. Coverage that is available is very expensive and typically not offered in areas that have a prior history of landslides. Most insurance policies are very specific and exclude any type of land movement, whether it's a landslide, a mudflow, or an earthquake. The Federal Emergency Management Agency may get involved with assistance, however, in cases where the area is declared a national disaster.

January 1982 Santa Cruz Mountains and Greater
San Francisco Bay Area

Between January 3 and 5, 1982, a catastrophic rainstorm, now known as an atmospheric river, dumped roughly one-third to half of the mean annual rainfall within a period of about 32 hours over the 10 counties in the greater San Francisco Bay area, triggering floods and landslides throughout the region. There were more than 18,000 debris flows that swept down hillsides in drainages, with virtually no warning, damaging at least 100 homes and killing 14 residents. From Marin County on the north to Santa Cruz County on the south and inland to Napa, Solano, Contra Costa, Alameda, and Santa Clara Counties, thousands of residents left their homes in threatened areas. Entire neighborhoods were cut off as roads were blocked, water systems were destroyed, and power and telephone connections were severed. The January storm damaged or destroyed 6,300 homes and 1,500 businesses, and also damaged miles of roads, bridges, and communication lines.[5]

The greatest amount of damage caused by this storm was from the thousands of debris flows. While the hillsides around the central coast were long recognized as being sites of historic landslides and other types of slope failures prior to this event, the potential for debris flows or mudflows was not appreciated. Nothing like the January 1982 storm had occurred and been mapped or recognized in the preceding decades. In addition, with the mostly densely vegetated slopes and high seasonal rainfall in the mountainous areas, native vegetation quickly regrew on the scars from previous debris flows, which made them nearly impossible to recognize, even by geologists. Attention had been focused on the larger, deeper, usually slower-moving landslides and slumps, rather than these smaller but potentially more dangerous and deadly rapidly moving mud/debris flows.

Late in the afternoon of Sunday, January 3, 1982, a light rain began to fall in the upper reaches of the San Lorenzo Valley in the Santa Cruz Mountains. The rainstorm, coming out of the southwest, rapidly increased to monsoon intensities and sustained that level for about 30 hours until late Monday, clearing just before midnight. The small mountain communities of Boulder Creek, Ben Lomond, and Lompico represented "ground zero" for this particular storm. Rainfall in these areas totaled 12 to over 15 inches during the 32-hour storm period, about one-third of the average annual precipitation, and equivalent to a 100-year storm.

Many residents who left for work Monday morning were unable to return home that evening, as much of the San Lorenzo Valley was cut off by either swollen rivers and creeks or the landslides, mud, and fallen trees that covered most of the mountain roads. North of Ben Lomond along Love Creek, the stranded residents of Love Creek Heights were set to weather the storm. The nearly 40 homes scattered across the hillside above Love Creek ranged from older summer cabins to modern homes. During the evening hours of January 4, the residents kept busy trying to stay warm and dry and maintaining the drainage systems and culverts around their homes.

The storm ended abruptly around midnight on January 4, and the sky was filled with stars. The only indication that a storm had recently deluged the area was the roar of Love Creek at the bottom of the canyon, raging out of its banks and tearing out the roadway. In the early moments of January 5, most of Love Creek's residents were asleep. And then at about 1:00 a.m., a 1,000-foot-long slab of the steep but rain-weakened hillside above them broke loose and rapidly swept more than 600,000 cubic yards of rock, mud, and debris down through Love Creek Heights towards Love Creek (see figure 8.1). This was the largest and most devastating landslide to ever impact the Santa Cruz Mountains and the central coast. While the Big Sur area has repeatedly been hit by larger slides (the Mud Creek slide of May 2017, for example, involved an estimated 2.4 million cubic yards of material), rarely have lives been lost in this sparsely populated area.

In its path, the 35-foot-thick flow of rock and earth (the equivalent of 60,000 dump truck loads) above Love Creek buried 12 houses and 10 people. Sadly, some never woke again. Most of those who died were apparently caught asleep in their beds, attesting to the rapid rate of the slide's descent into the canyon. Four others barely escaped with their lives and were rescued from the mass of landslide rubble and splintered homes the next day. Search parties were later to recover six bodies as they dug into the mud at the base of the landslide, but four persons were never found. On the basis of loss of life, this was the most destructive, naturally induced landslide at the time in California's history, and the third most tragic in western North America in this century.[6]

1983–2021 Big Sur, Highway One

The most landslide-prone road on the central coast of California is the Big Sur Highway, State Route One. This roughly 70-mile section of

FIGURE 8.10. A moderate-size debris flow closed Highway One during the very wet winter of 2022–23. *Courtesy of Caltrans District 5, public domain.*

coast from Soberanes Point on the north to Ragged Point on the south was essentially isolated and not accessible by any vehicles until the highway was originally completed in 1937. Building this road was difficult from start to finish, and keeping this highway open remains a yearly challenge (figure 8.10). The mountains are among the steepest anywhere along California's coast, and the rocks are notoriously weak and prone to failure, particularly during very wet winters when the area may receive up to 80 inches of rain. Caltrans has named most of the large slides (Pitkins Curve, Hurricane Point, Grayslip, Grandpa's Elbow, Big Slide, and Mud Creek, to name a few) simply because they have failed repeatedly, which has required regular removal of the debris from landslides and slumps in wet winters. With several of the larger slides, completely rebuilding the highway has been needed. Highway One has been closed for months at a time as the process of rock and debris removal and reconstruction takes place.

A 1983 landslide near Julia Pfeiffer Burns State Park led to the closure of the highway for over a year, with repair work costing about $20 million (in 2022 dollars). The wettest winter in recent California history

(2016–17) again led to some very large landslides, including the massive Mud Creek slide, which involved 6 million cubic yards of rock and soil and closed Highway One for over a year, with road reconstruction costs exceeding $50 million (see figure 8.2). Repairing and rebuilding the highway for a landslide of this magnitude is a major undertaking and requires a lot of heavy equipment. The Mud Creek slide actually pushed the shoreline seaward a considerable distance, but because of the loose nature of the material involved, waves began to erode this rock and soil, which would have led to the destabilization of the regraded area where the road was being rebuilt. The decision was made to armor the base of the slide with large rocks brought in from near Cambria, about 35 miles to the south.

Beginning on August 18, 2020, the Dolan Fire rapidly spread and ultimately burned over 132,000 acres (~206 square miles) of steep terrain in the mountains of Big Sur. In late January 2021, an atmospheric river hit California and dumped 10 to 13 inches of rain on the area burned over the prior summer. With much of the vegetation having been burned off, rainfall quickly turned to runoff, which came cascading down the steep drainages, carrying mud, logs, and other debris down towards Highway One. The Rat Creek drainage ended in a culvert under the highway but was totally incapable of handling the water and debris that was delivered. The debris quickly plugged the culvert and began to flow over the roadway and erode the fill supporting the highway. One hundred feet of roadway was carried away, which closed this important corridor for nearly three months (see figure 1.1).

1995 and 2005 La Conchita, Ventura County

La Conchita is a small neighborhood of about 200 single-family homes on the coast 10 miles north of the city of Ventura. The homes sit on a narrow coastal strip about 800 feet wide between the shoreline and steep 600-foot-high bluffs. This small community has a long history and originally was home to the workers who built the railroad along the coast here in the latter half of the 19th century. In later years the residents were typically the laborers working in the oil fields.

The cliffs immediately inland from La Conchita are backed by a high terrace that has been intensively cultivated with avocado and citrus orchards. The bluffs above La Conchita consist of poorly consolidated marine sediments, primarily shale, mudstone, siltstone, and sandstone. These weak rocks have a long history of failure through landsliding.

Historic accounts dating back to 1865 reported frequent landslides in the area. Slides inundated the Southern Pacific railway in 1889 and again in 1909, when a train was actually buried by debris.

On March 4, 1995, at 2:30 in the afternoon, the bluffs above the small community failed with no warning, and hundreds of cubic yards of material moved down the steep slope in just a few minutes. While this failure began as a slump, it developed quickly into a more fluid earthflow, either seriously damaging or destroying nine houses. This slump-earthflow was large, 400 feet wide, 1,100 feet long, and covered about 10 acres. The depth of the slide was estimated at about 100 feet, and the overall volume at 1.7 million cubic yards. That amount of material would fill about 170,000 dump trucks. Six days later on March 10, a debris flow from a canyon to the northwest damaged five more houses in La Conchita.

This large slope failure was believed to have been initiated by a period of very high precipitation. Rainfall records from the closest precipitation gauge (12 miles away in Ojai) indicate that between October and the day of the slump-earthflow, rainfall was nearly 30 inches, about twice the long-term average. Most of that rain (24.5 inches) fell in January 1995, a month that typically only sees about 4.3 inches. It took some weeks for the rainfall to percolate deep into the slope to the point of failure. Fortunately, despite the magnitude of the failure, there were no injuries or fatalities.

However, almost exactly 10 years later, on January 10, 2005, about 250,000 cubic yards of the southeastern portion of the 1995 slump was remobilized and flowed into the neighborhood closest to the base of the hillside (figure 8.11). Thirteen houses were destroyed, 23 more were damaged, and there were 10 fatalities, residents who sadly were buried in their homes. A retaining wall built at the base of the slope following the 1995 earthflow was tilted forward or was overtopped by the 2005 flow. That same day, rapidly moving debris flows from canyons immediately to the northwest of La Conchita flowed onto and covered Highway 101. These 2005 flows were also likely initiated by prolonged heavy rainfall. Between December 27, 2004, and the day of the failure (January 10, 2005), the city of Ventura just 10 miles east recorded 14.9 inches of rainfall, nearly as much as its average annual precipitation (15.4 inches).

Because of its setting, at the base of steep and unstable slopes, the La Conchita community will be prone to future slope failures, particularly under prolonged or sustained rainfall. Historic slump materials can again be mobilized under these conditions, or new slope failures can be initiated. Virtually the entire La Conchita neighborhood is exposed to

FIGURE 8.11. In the 2005 La Conchita landslide-earthflow, the fluid portion reached Highway 101, which temporarily was closed. *Courtesy of US Geological Survey, public domain.*

these future hazards, and there is no easy way to predict when failures may occur. Continued monitoring of rainfall is the best approach for determining when slopes are likely to be activated and when evacuation should be undertaken or considered.

1998 Laguna Niguel

Southern California had one of the rainiest Februarys in its history in 1998, when continuing storms late in the month delivered as much as 13.5 inches of rain, the highest monthly total in the previous 30 years. This led to conditions in May that generated 56 major landslides with seven fatalities. In March 1998, the Niguel Summit neighborhood in the city of Laguna Beach was shocked by the Via Estoril landslide, which destroyed nine homes and damaged portions of the Crown Cove condominium complex, which was ultimately demolished before additional damage could occur. In 1985, prior to the construction of this development, the developer was told by his geological consultant that the 900-acre site was underlain by six landslides and that the stability of the ground was "generally less than acceptable" and that large areas were "probably unstable." Despite this warning, the developer moved forward, and the responsible local government approved the project. By the 1990s, some of the houses constructed in the area were starting to develop cracks in their walls and also in their driveways and sidewalks.

The legal settlement money from the destructive 1998 landslide was used to build a large concrete retaining wall with tieback anchors, a buttress, and subdrains. At the time, local officials explained to the residents that it was highly unlikely that the area would ever be built on again, although it continued to be zoned for housing.

Having lived through the 1998 event and seen the destruction firsthand, the residents of the Laguna Niguel neighborhood, with their views and swimming pools, were quite certain that this area would never be built on again. Twenty-four years later, however, in 2022, the original developer who had built about 1,500 houses in the area, including the original Niguel Summit neighborhood homes and condominiums, proposed developing the area immediately below the old landslide site with a new 22-townhome project. The proposal called for removal of part of the soil buttress at the base of the prior slide, in order to provide space for a retaining wall and yards for the proposed townhouses. With the astronomical price of homes in many Southern California coastal communities, development and redevelopment has often taken place in unstable areas with the argument or case made that the site is now stable. Stability of hillside land in many places is very difficult to guarantee, however, as past experience has shown.

January 9, 2018, Montecito Debris Flows

Like many of the arid chaparral-covered hillsides of the Santa Monica and San Gabriel Mountains of Southern California, the Santa Ynez Mountains above Santa Barbara and Montecito have also experienced devastating fires, which seem to be occurring more frequently and covering larger areas with a changing climate. On December 4, 2017, the Thomas Fire, which erupted near Santa Paula, some 40 miles southeast of Montecito, was to wreak havoc and burn over 280,000 acres, or 438 square miles, of dry mountainous terrain. The blaze burned for over six months, destroying over 1,000 structures and causing an estimated $2.6 billion in damage (in 2022 dollars). The fire literally scorched the earth, burning off the cover of trees and brush. With the vegetation now gone, and no protection for the soils, the burned and steep slopes were extremely susceptible to any sustained or high-intensity rainfall. The heat from the fire incinerated the surface layer of organic matter that would normally absorb and hold the moisture. Instead, an essentially water-repellent layer was created that led to rapid runoff of the short-duration but intense rainfall of the early morning of January 9.

Just half an inch of rain fell on the hillsides above Montecito in about five minutes at 3:30 a.m. on January 9, 2018. The water from this intense rainfall quickly began to flow downhill, picking up the soil and turning it into mud. As the flow increased in volume and velocity, it picked up cobbles, then boulders, and soon turned into a destructive debris flow.

While the mud and debris were initially concentrated in a few stream channels, the increasing volume of water and debris from upslope, as well as what was being picked up along the way, soon overtopped the channels and spread out on either side. While no one was measuring all of this in the darkness of the early morning hour, the mud marks and debris left behind indicate the flows were up to 15 feet thick and were probably moving about 20 miles per hour. This is considerably faster than the average person can run.

Twenty-three people died and 65 houses and eight businesses were destroyed as the mud and debris flowed rapidly through a swath of Montecito with very little warning; 163 more people were hospitalized, and 462 more homes and 20 more businesses were damaged (figure 8.12). While there apparently were evacuation warnings, it seems that after the multiple fire evacuations in December, many people were simply suffering from evacuation fatigue and chose to stay at home. Some were fortunate and survived, and some sadly were not, even within the same home. Cars and trucks were picked up like toys by the debris flows, twisted and squashed, and two were carried all the way to the beach, as was a stuffed bear, apparently a hunting trophy swept out of someone's home. Some houses were ripped off their foundations while others were completely filled with mud and boulders.

The power of thousands of cubic yards of mud and debris, 5 to 15 feet thick, moving at 20 miles per hour through a normally quiet, semi-rural neighborhood is truly hard to imagine until you see the photographs of crushed cars and of homes reduced to piles of rubble. Much of the mud and debris came to rest on Highway 101, completely closing it off to traffic for at least two weeks.

FIGURE 8.12. Destruction of a home and vehicle from the 2018 Montecito debris flows. *Courtesy of the California Governor's Office of Emergency Services, public domain.*

December 2022–January 2023 Rainfall and Slope Failures

The nine atmospheric rivers that drenched California between December 26, 2022, and January 17, 2023, dropped 20 to over 30 inches of rain in many of the state's mountainous areas, which led to at least 300 slope failures scattered from north to south and east to west. California's youthful geology, steep slopes, and often unstable soils are prone to mud and debris flows after prolonged and/or high-intensity rainfall, particularly after fires have removed the protective vegetation cover. Shallow and rapidly moving mud and debris flows and slides responded quickly to the sustained and intense rainfall of early January, and these failures blocked roads and state highways in mountainous areas across the state. Prolonged rainfall in the intensely developed hills of Oakland and Berkeley also generated some earth and mud flows that damaged homes.

While mud and debris flows and slides typically occur soon after the soils have been saturated, it usually takes longer for prolonged precipitation to infiltrate deeper into the subsurface and destabilize a slope to create a deep-seated landslide or slump. These conditions did occur in

January in a few places, and landslides or slumps impacted several stretches of highways in the Coast Ranges.

It is tempting to draw comparisons between the January 3–5, 1982, storm and the landslides, debris flows, and other slope failures associated with the December 26, 2022–January 17, 2023, atmospheric rivers. While both winters were damaging to large parts of the state, there were some significant differences in the areas impacted and the extent of slope failures. The January 1982 event was a single storm that parked over the seven counties of the greater San Francisco Bay area for about 32 hours and dumped roughly a third to half of the mean annual rainfall in those three days. This storm followed well-above-average rainfall in November and December of 1981, so the hillsides were primed for failure. When the early January rains began to fall, the soils were already nearly saturated, so additional rainfall pushed them over the edge. Saturation turned the soils and overburden in many swales and steep drainages into a liquid state, and about 18,000 debris flows ensued (see the earlier section in this chapter on the January 1982 winter). Many lives were lost and many homes were destroyed or damaged, with the Love Creek slide and debris flow being the most damaging and destructive.

The atmospheric rivers of 2022–23 spread out the precipitation over about 23 days instead of 32 hours, and they also fell at a time when California was in a drought condition. About 80 percent of the state was designated in early December as being in an exceptional, extreme, or severe drought. Nonetheless, the 20 to over 30 inches of rain that descended in the coastal mountains did weaken the soils on the hillsides and roadcuts, and as of late January 2023 had generated at least 300 slope failures. Most of these were earth and debris flows, rather than deeper-seated landslides, and many of these occurred along oversteepened roadcuts. There is always some uncertainty involved in grading roads through steep topography. The steeper the cuts are, the less grading is involved, which means lower construction costs and less land that has to be acquired; on the other hand, all other things being equal, steeper slopes are less stable, especially when saturated or under prolonged or high-intensity rainfall.

SOME FINAL THOUGHTS ON SLOPE FAILURES

In a geologically young and active terrain like the mountains and hillsides of California, the weak nature of many of the underlying rocks, the often-steep slopes, the high potential for intense or prolonged winter rainfall—which is likely to become more concentrated with ongoing climate change—and the seismically active nature of the region combine to produce a landslide or debris flow–prone landscape. Adding to the natural factors producing instability are the many human actions that have sculpted and disturbed the natural terrain of the hillsides and mountainous areas, especially surrounding the major urban centers such as the greater Los Angeles and San Francisco Bay areas. Development has involved grading for highways, streets, and building sites; adding excess water from septic tanks and leach fields; and runoff from increased impermeable surfaces such as roofs and roads, sidewalks and parking lots, to name the most obvious. These landscape modifications have affected the stability of the landscape, usually in negative ways. And even without human modifications, landslides, slumps, debris flows, and earthflows are parts of a natural process of landscape evolution and only become problems or hazardous when we get in their way. Hillsides will continue to fail in California, and prolonged or high-intensity rainfall will likely increase slope instability. Experienced geologists can usually recognize landslides and landslide-prone terrain, but the winter of 1982 and the 18,000 mud and debris flows also surprised and humbled even the best field geologists, who simply had never experienced slope failure on this scale.

Nature, to be commanded, must be obeyed.
—Francis Bacon

9

Coastal Storms, Sea-Level Rise, and Shoreline Retreat

INTRODUCTION

Although California's 19 coastal counties only account for 22 percent of the state's land area, they are home to 66 percent of its people, 80 percent of its wages, and 80 percent of the state's GDP. That coastal population is not distributed evenly, however, with three Southern California counties alone (San Diego, Orange, and Los Angeles) home to 41.6 percent of the state's residents (16.3 million people) and those counties around San Francisco Bay containing another 17.8 percent of California's population.

Coasts, however, are some of the most dynamic environments on the planet, with nearly constant change all but guaranteed. Beaches come and go seasonally, dunes advance and retreat, and cliff and bluffs fail and erode on a recurring basis. Yet this is where many Californians have chosen to live, with those homes closest to the edge—whether on cliffs, bluffs, beaches, or dunes—commanding the highest prices, but also exposed to the greatest risks. This wasn't the case a century ago, when the earliest arrivals didn't have the same magnetic attraction to the shoreline that the residents of today do. There seems to have been either more awareness of the risks inherent in living directly on the shoreline, or just less interest in living literally on the edge.

I live 500 feet from the edge of the Pacific Ocean—close enough to hear the waves breaking during a major storm but far enough away to feel safe and unthreatened. The bluff is just a three-minute walk away, and it's one I take frequently. On the morning of January 5, 2023, I could hear the pounding of the surf on the shoreline and also knew it was a very high tide—the coincidence of very large waves and a very high tide is historically when most of the coastal erosion and wave impacts have occurred along California's coast. It was still early, maybe 7:00 a.m., but I got my camera and headed for the street that follows the bluff edge. In my 55 years in Santa Cruz, chasing storms and observing damage, I had never seen the sea this angry. Waves were overtopping the 20- to 30-foot-high bluffs, and water was crossing the road. The few cars out were getting splashed with seawater, and city police had just closed the road. Within a few hours at high tide, the waves had chewed into the sidewalk and were working their way into the roadway (figure 9.1). This storm was to shape how the city was going to deal with coastal erosion for years to come.

We have learned in the last 20 or so years that the climate in the Pacific Ocean and along the California coast oscillates over periods of several decades, between warmer and cooler intervals termed Pacific Decadal Oscillations (PDOs, figure 9.2). The warmer or positive periods are characterized by higher ocean surface water temperatures with greater evaporation and subsequent rainfall, which can generate flooding and landslides in coastal regions. During these periods we also experience more frequent and severe El Niño events, with coastal storms and more energetic and damaging waves along the coastline, typically approaching from a more westerly or southwesterly direction.[1]

The period from about 1926 to 1944 was dominated by these warmer ocean surface conditions and a generally stormier coastal climate. About 1945, however, the PDO interval transitioned to a generally cooler and calmer period. This coincided with the end of World War II, when California's population expanded rapidly. The state's population grew from 9.34 million in 1945 to 22.35 million in 1977, a 240 percent increase. In order to meet the housing needs of all those new residents, the home construction industry boomed. Like the new housing tracts, the coast was also subdivided, and homes were built on what appeared at the time to be stable and safe cliffs, bluffs, sand dunes, and even the beaches themselves. Those ocean view homes and properties became increasingly valuable in subsequent years.

About 1978, however, the PDO switched rather abruptly back to a warm, stormier cycle with more frequent and more powerful El Niño events. All of those homes that had been built on the coast or along the

FIGURE 9.1. Storm waves at a very high tide in early January 2023 eroded the coastal bluff and chewed away at the sidewalk and street along West Cliff Drive in Santa Cruz. © *2023 Gary Griggs.*

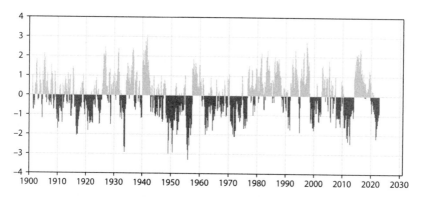

FIGURE 9.2. Pacific Decadal Oscillation (PDO) cycles from 1900 to 2022. The vertical axis is Pacific Ocean sea surface temperature anomaly in degrees Celsius. *Modified from National Centers for Environmental Information / NOAA.*

shoreline were now threatened by elevated sea levels that accompany the warmer water of an El Niño, as well as larger storm waves, typically arriving from the west and southwest, rather than the more frequent arrival from the northwest. Damaging winters along the shoreline arrived in 1978, 1982–83, and 1997–98, which quickly changed perceptions about the wisdom and risks of living on the edge.

COASTAL PROCESSES AND HAZARDS—SHORT TERM AND LONG TERM

Short of the crater rim of an active volcano, there are few places on Earth that are as continuously active as the shoreline. If you've ever watched winter storm waves battering a coastal bluff (figure 9.3), or very high tides and large waves sweeping across the beach and onto roadways, through parking lots, or into living rooms (figure 9.4), you can appreciate the tremendous power of the ocean. It has usually been the simultaneous occurrence of very large winter waves and the highest monthly tides that has produced the greatest shoreline erosion and damage to either private property or public infrastructure along the state's coast.

Short-Term Hazards and Events

Hazardous events along the coast or shoreline are quite different from those caused by large earthquakes, tsunamis, or floods. These last three events can be devasting and deadly due to both their magnitude and the size of the areas affected. Other than very large but relatively infrequent subduction zone earthquakes and their associated tsunamis—covered in chapters 2 and 3—the processes that affect the coastal zone are not usually catastrophic events with large-scale death and destruction. There are the day-to-day impacts of waves breaking on the shoreline—waves that change in height and energy seasonally—leading to beach accretion in the spring and summer months and erosion in the winter months, which gradually wear away the coastal bluffs and cliffs. Then there are the short-term, large storm waves, which have greater impacts on beaches and bluffs and can be more destructive to developed shorelines, particularly if they arrive at times of very high tides or elevated water levels. In addition, there is the long-term, gradual but accelerating rise in sea level, which will impact the developed California shoreline for decades and centuries to come.

FIGURE 9.3. Winter storm waves in January 2023 battering the coastal bluff in Santa Cruz. © *2023 Gary Griggs.*

FIGURE 9.4. Very high tide washing across East Cliff Drive in Santa Cruz. © *2023 Micheal W. Beck.*

WHAT IS AN EL NIÑO, AND HOW DOES IT AFFECT CALIFORNIA?

Peruvians have known for at least four centuries that the intrusion of a current of warm water from the western equatorial Pacific every several years leads to a dramatic reduction in the population of anchovies in coastal waters. This anchovy decline not only severely impacts their fishing industry, but also causes a decline in marine mammals and sea birds that rely on anchovies as a food source. In addition to the ecological impacts, this intrusion of warm water produces higher evaporation rates that typically lead to torrential rainfall on the adjacent continents, accompanied by floods, mudflows, and landslides. Although the extra rainfall has a positive effect on plant growth in an otherwise dry coastal area, the large-scale negative consequences of this event over time have long been recognized.

In Peru, because this warm-water phenomenon often arrived around Christmastime, it was given the name El Niño, Spanish for "the child," in reference to the birth of Jesus. More recently, particularly over the past 45 years, the global scale of this event and its widespread and diverse impacts have been more fully understood. Scientists came to understand that an El Niño event is linked to an alteration of atmospheric pressure systems over the equatorial Pacific. They called this alteration the Southern Oscillation. This large-scale phenomenon is now more accurately called El Niño–Southern Oscillation, or ENSO, and is characterized by major, interconnected shifts in both atmospheric and oceanic circulation throughout the entire Pacific basin.

In what are thought of as "normal" times, trade winds blow from east to west, both north and south of the equator, and move warm surface waters toward the western equatorial Pacific, where a large pool of warm water accumulates. Every three to seven years, however, the normal atmospheric circulation system over the Pacific Ocean breaks down for reasons that are still not understood. The trade winds weaken, atmospheric pressures across the Pacific reverse, and winds actually begin to blow from west to east, moving the warm pool of water gradually back toward the coast of South America, initiating an El Niño. The size of the pool of warm water is usually a good indicator of the size of the emerging ENSO event.[2]

Through the use of satellites and moored surface buoys, we can now accurately determine and monitor the size, temperature, and movement of this pool of warm water. It was because of this monitoring that warnings of the large 1997–98 El Niño event were circulated months in advance by NOAA and the National Weather Service. There is, however, a long time period between the initial oceanographic observations in the spring that suggest an El Niño building in the western Pacific, and its ultimate arrival on the coast of California, usually in the winter. Although the 1997–98 ENSO event was predicted quite accurately, the impact wasn't really felt until January through March of 1998, almost a year after the initial predictions. As a result, many coastal residents concluded that this event simply wasn't going to take place as announced, and some newspaper editorials even proclaimed the event as "El No Show." It is important to realize, however, that whether a large El Niño or a major earthquake, there is a limited number of things that we can do to reduce the impacts of such an event even when it has been predicted. Certainly lives can be protected through warnings and evacuations, but homes, businesses, roads, and other infrastructure cannot be relocated.

When the warmer equatorial water reaches the coast near Ecuador and Peru, it splits and moves both north and south along the coast, raising sea levels along the shoreline and transporting normally tropical species northward into California waters. The effects of ENSO events on the coastline of California were widely unappreciated, not fully understood, or simply ignored by many permitting agencies, coastal builders, developers, real estate agents, and homebuyers in the past. The El Niño winters of 1977–78, 1982–83, and 1997–98 inflicted major damage on many of the developed and heavily populated portions of California's coast and changed that perception, however, by bringing an end to the relatively benign climate of the previous three decades of a cooler and calmer PDO interval (see figure 9.2).[3]

Coastal erosion and storm damage along the coast of California are maximized when several factors or processes occur simultaneously. Historical research focused on the central coast of California has shown that about 75 percent of the storms that caused significant erosion or damage over the past century occurred during El Niño winters. The associated factors that contributed to the severe coastal erosion and destruction during these events included sea levels that were higher than normal or predicted, as well as more frequent and larger storm waves arriving from the west or southwest coincident with high tides.

The 1982–83 El Niño

During the first three months of 1983, higher sea levels, larger and more frequent waves, and high tides converged to inflict the greatest amount of documented shoreline erosion and coastal storm damage the state's coastline had experienced in the previous 50 years. The El Niño brought a pulse of warm water from the tropics northward along the California coast, which raised sea levels above predicted high tides by up to two feet. These were the highest water levels recorded in all the historic tide gauge records at San Diego, Los Angeles, and San Francisco. Large storm waves approached the coast from the west and southwest during these three months. As a result, there was very little energy loss through bending or refraction, as usually happens with the more typical waves from the northwest. Consequently, the full force of the waves struck much of the state's coastline and the associated development and infra-structure. The third factor, the occurrence of very high tides (up to and exceeding six feet MLLW, or mean lower low water), coincided with the arrival of the large waves from seven different storms. The elevated sea levels and large waves damaged breakwaters, piers, park facilities, seawalls, coastal infrastructure, and public and private structures. Win-ter storm damage along the coast of California totaled over $300 mil-lion (in 2022 dollars). Thirty-three oceanfront homes were completely destroyed, and about 3,000 homes and businesses were damaged (figure 9.5). Oceanfront public recreational facilities suffered an estimated $91 million in damage (in 2022 dollars).[4]

Along the central coast, wave impact, high tide flooding, sea cliff ero-sion, and undermining of coastal structures occurred from Stinson

FIGURE 9.5. Storm damage in 1983 along Via Gaviota at Aptos Seascape, northern Monterey Bay, where waves at high tide overtopped the protective rock revetment that ended up serving as a ramp. © 1983 Gary Griggs.

Beach to central Monterey Bay. Over 65 feet of bluff recession in Pacifica forced the removal of an entire row of mobile homes in a large ocean-front mobile home park and also a two-story building. This area of high but weak bluffs was to become an epicenter for continued erosion and loss or demolition of bluff-top homes and apartments in the subsequent decades.

In northern Monterey Bay, private homes, restaurants, seawalls and bulkheads, a state recreational vehicle campground and picnic area, sewer lines, and roads were severely damaged. At Seacliff State Beach, which is a popular back-beach RV camping and picnic area, a number of historic winter storms had repeatedly destroyed major sections of the timber bulkhead/seawall in the past, and the 2,670-foot-long wooden structure had been rebuilt again in late 1982, at a cost of $13.7 million (in 2022 dollars), with the expectation that it would last 20 years. But in late January 1983, within two months of its completion, waves, high tides, and large logs battered down 700 feet of the new structure. Eleven RV sites were destroyed, restrooms were heavily damaged, and logs and debris were washed back to the old sea cliff, cutting off access to ocean-front homes immediately upcoast. Losses were estimated at $5,400,000 (in 2022 dollars). This was the seventh time this bulkhead had been built, damaged or destroyed, and then rebuilt.[5]

Immediately downcoast from Seacliff State Beach, at the south end of Beach Drive in Rio Del Mar, 26 beach-level homes were thought to be protected by a variety of timber and concrete seawalls. These were pro-gressively battered, broken, or undermined and failed. Virtually every protective structure was damaged or destroyed. Two houses with shallow foundations were total losses, while others lost pilings, windows, decks, and stairways. The sewer main beneath the beach serving the beach-level homes was exposed and severed, followed by two weeks of raw sewage release. Nine hundred feet farther southeast at the Aptos Seascape back-beach development, large waves overtopped the riprap fronting 21 homes. Nineteen of the homes suffered damage as waves broke through win-dows, sliding glass doors, and house fronts (see figure 9.5).

Major dune erosion threatened homes at Pajaro Dunes west of Watsonville at the mouth of the Pajaro River. Luxury vacation homes had been built on the frontal dune, which is a very ephemeral feature and subject to erosion and retreat during periods of large waves and high tides. Erosion took place very quickly, and continued wave action threatened at least 17 homes and 25 condominiums. By late January up to 40 feet of dune had been eroded, leaving a vertical scarp up to 18 feet high adjacent to the foundations. Emergency riprap, brought in and placed in front of at least 60 homes, was later replaced by a more permanent revetment.

The statewide damage from the 1982–83 ENSO event was a wake-up call for coastal California and, at least in the short term, changed popular perceptions of the stability or security of oceanfront development. Development restrictions were tightened, many homes went on the market, insurance rates rose, and many requests for new seawall permits were submitted. Then it was back to business as usual, as new arrivals to the coast outbid each other for the same oceanfront homes damaged a few years before.

Any evaluation of coastal erosion rates or the long-term stability of an oceanfront property in California must factor in the ENSO cycles and the decadal oscillations (see figure 9.2) that are now well documented. The record of coastal change in a given location for a particular interval of time, 1945–78 for example, an interval characterized by relatively few and modest El Niño events, may seem to encompass a long time frame but may not be at all representative of the longer-term conditions that the site has experienced or may experience during an interval dominated by ENSO conditions.

The 1997–98 ENSO Events

Significant coastal storm erosion and damage also occurred in the 1997–98 winter, another large ENSO event. Although most indices or indicators suggest that the 1997–98 ENSO disturbance was more intense than the 1982–83 event, the state's coastline suffered far more damage in the earlier winter. Some important differences have been documented that explain why the earlier event was so much more damaging and why property losses were greater. During the first three months of 1983, eight major storms struck the coast. The two largest storms struck during high spring tides, so the waves scoured away more beach sand, reached farther inland, and damaged structures higher on the shoreline. In 1997–98, the largest waves from the two biggest storms hit during lower periods in the monthly tidal cycles, significantly reducing the impact of the waves on the shoreline. Nonetheless, there were 17 storm-related deaths in California, and at least 27 homes were red tagged as unsafe to occupy within the coastal zone in the 1997–98 winter.

Another important factor contributing to the disproportion of damage during the two winters was the difference in percentage of shoreline that had been armored. Most of the areas significantly damaged in the 1982–83 winter were protected by more substantial seawalls or revetments by the time the 1997–98 ENSO event arrived.

The warmer, stormier PDO interval that began in 1977 shifted to a generally calmer and cooler period following the destructive 1997–98 El Niño, and the calm lasted about 16 years. The transition back to an El Niño–dominated phase was abrupt, with the 2015–16 El Niño being one of the three strongest events ever recorded—the other two being in 1982–83 and 1997–98. Most beaches along the California coast eroded beyond historical extremes in the 2015–16 event. Little damage to development was reported; in large part this is believed to have been due to the level of coastal armoring emplaced in the previous decades.

January 2023 Winter

Although there have been several large ENSOs in the years since the damaging 1982–83 event, it wasn't until January of 2023, which somewhat strangely wasn't an El Niño year, when the simultaneous occurrence of extraordinarily large storm waves, very high tides, and strong onshore winds produced damage comparable to the 1982–83 winter, particularly along the central coast. Waves again arrived from the westsouthwest, so hit the coast head-on. A wave buoy offshore Monterey Bay recorded significant wave heights, up to 28 feet on January 5 and 6 (see figure 9.3), even larger than in the early 1983 storms, and these waves arrived coincident with very high tides, which were precisely the conditions that led to the coastal damage during the first three months of 1983. Strong onshore winds added to the impacts.

Not surprisingly, the Monterey Bay area sites most heavily impacted in early January 2023 were very similar to the areas affected by the waves and high tides of 1983. These included the low-lying city of Capitola, which has repeatedly been damaged by storm waves throughout its history. Restaurants along the shoreline esplanade and condominiums at beach level, as well as an old pier that also had been ravaged by historic storms, were all seriously damaged.

At Rio del Mar, in the northern corner of Monterey Bay, the ocean overtopped seawalls, flowed into downtown, and brought sand and debris, including large logs, up against and into some homes built on the beach. Any structure built on the sand or at beach level, while providing wonderful views and easy access to the beach, is sooner or later going to be flooded by high tides and storm waves. Sea-level rise will

FIGURE 9.6. The Seacliff State Beach timber bulkhead that supported an RV campground was destroyed by waves, logs, and debris for the eighth time in 2023. © *2023 Kim Steinhardt.*

only exacerbate this hazard. Immediately upcoast of Rio Del Mar, the Seacliff State Beach timber bulkhead that supports an RV campground, parking lot, and picnicking spot was destroyed for the eighth time since first constructed in 1926 as logs and trees broke through the timbers, followed by erosion of the fill behind the bulkhead (figure 9.6).

Although most of the damaged areas were low-lying, the large waves arriving at high tide also overtopped the 20- to 30-foot-high bluffs along the oceanfront West Cliff Drive area of the city of Santa Cruz. Protective riprap collapsed, and a pedestrian and bike path and part of the roadway were destroyed at several locations (see figure 9.1).

COASTAL LANDFORMS AND RISKS TO DEVELOPMENT

It would be difficult to find a coastline anywhere in the world that has had a more complex geological history than California's, and it is because of the vastly differing rock types and regional geological histories that specific sections of the state's coastline look and evolve so differently. Some rocks are so hard and resistant to erosion that photographs taken of the coast 50 or more years ago look identical to those of today. Elsewhere, however, coastal bluff materials are so weak and erodible that the coast is retreating at rates of five feet or more each year. These changes are usually easily recognized by comparing historical photographs.

Although large-scale coastal landforms such as mountains, uplifted marine terraces, and sea cliffs owe their relief and origin to regional tectonic events, surface processes such as wave attack have subsequently altered them all. In addition, sea level along the coast has changed continuously through time such that the position of the shoreline at any particular time is only a temporary one. Although these changes are not rapid, it is evident from historical photographs and geological evidence that the shoreline undergoes constant change. There is every reason to believe that the processes that have shaped the coastline in the past will continue to reshape it in the future.

THE DISTRIBUTION OF COASTAL LANDFORMS

One way to help understand why particular sections of California's coastline look so different and behave so differently is to break down the state's coast into some distinct and recognizable landforms, each of which presents different hazards and risks, either to homes and businesses or to public infrastructure. One straightforward breakdown is to identify any particular part of the coast as consisting of (1) steep coastal mountains and sea cliffs, typically with hundreds of feet of relief; (2) uplifted nearly horizontal marine terraces and lower sea cliffs and bluffs varying from 20 to perhaps 200 feet in height; or (3) low-relief shoreline areas with beaches, sand dunes, estuaries, or lagoons. As soon as we make up a coastal classification or a series of pigeonholes, however, we almost immediately recognize that there are transitional or intermediate areas that fall between the others, and that's fine. Nonetheless, beginning to recognize these distinct landforms helps us to understand how and why different parts of the coast look and act like they do today and why development and infrastructure in each of these environments will be exposed to distinct hazards and risks.

The great majority (72 percent, or about 790 miles) of the 1,100-mile California coast consists of actively eroding sea cliffs or coastal bluffs. Of this 790 miles, about 650 miles consist of lower-relief cliffs and bluffs typically eroded into raised marine terraces; the other 140 miles consist of high-relief cliffs and coastal mountains. The remaining 310 miles or 28 percent of the coastline is of low relief and relatively flat. These are the wide beaches and sand dunes, as well as the bays, estuaries, lagoons, and wetlands, that form many of the state's coastal recreational environments and parkland. Good examples include the entire Monterey Bay coastline, the Santa Monica Bay shoreline, and the coast between Long Beach and Newport Beach.

Steep Cliffs and Rugged Mountains

The state's high-relief, steep cliffs and coastal mountains are predominantly in Northern and Central California from Del Norte to Mendocino County, at the Marin Headlands just north of the Golden Gate, from Pacifica to Montara in San Mateo County, and along the Big Sur coast of Monterey and San Luis Obispo Counties. High-relief cliffed areas and rocky headlands also characterize the Santa Monica Mountains coast between Point Mugu and Malibu. Because these spectacular areas are so hard to access and develop, they provide some of the state's most pristine coastal scenery. Only limited development has taken place along these steep coastlines, and most of the mountainous areas are now protected within national forests and state parks.

These rugged stretches of coast typically consist of older and harder rocks and, as a result, are very resistant to erosion and form many of the protruding headlands or points along the coastline. These rock types generally erode very slowly but in some areas, such as the Big Sur coast, are subject to large-scale landsliding, slumping, and debris flows (see chapter 8).

The steep topography of the Big Sur coast presented a major obstacle to overland travel from the time of the earliest Spanish explorers in 1769 until Highway One was finally blasted through the solid rock in the late 1920s and 1930s. Today this highway still clings precariously to the rocky cliffs (see figure 8.2). This narrow ribbon of access is a white-knuckle driving experience that traverses an otherwise mostly virgin coastline rising 5,000 feet to the crest of the Santa Lucia Mountains. Maintaining this highway remains an ongoing challenge, and the road is frequently closed by landslides and rockfalls in the winter months. Following the heavy rainfall of the 1982 and 1983 winters, the Big Sur coast was closed for about eight months until massive amounts of rock and soil could be removed from the roadway, hillsides could be regraded and stabilized, and Highway One could be completely rebuilt.

On May 20, 2017, the steep mountainside at Mud Creek on the Big Sur coast, 65 miles south of Monterey, failed catastrophically and carried about 6 million cubic yards (roughly 600,000 dump truck loads) of rock and soil downslope, removing approximately 1,300 feet of Highway One and pushing the shoreline about 570 feet seaward (see figure 8.2). This wasn't the first time this area has failed, although past surveys indicated that this particular hillside was characterized by a slow downhill creeping motion, such that this catastrophic failure was somewhat unexpected. In this case, a five-year drought followed by continuing and saturating rainfall appears to have been the triggering factor for failure. Complete repair and reconstruction of the highway took 14 months, had a price tag of $54 million, and included stacking very large rocks at the base of the slide on the shoreline to build a rock revetment and protect the landslide material from wave erosion.

Each wet winter brings new hillside failures and road closures in this rugged coastal terrain, particularly if the mountainsides inland from the Big Sur Highway have been recently burned. The Dolan Fire burned from August 18 to December 31, 2020, and eventually scorched over 128,050 acres. This set the stage for rapid runoff during the intense rainstorm (designated an atmospheric river) that dropped as much as 15 inches in the Santa Lucia Mountains above Big Sur just a month later, between January 26 and 28, 2021. Burned hillsides tend to repel the rainfall rather than allowing it to percolate into the ground as would occur under normal or unburned conditions. The rapid runoff mobilizes soil, rock, and other debris that has accumulated on the slope, which then flow rapidly down the steep slopes and can cause damage and deadly devastation downslope.

A large portion of the Rat Creek watershed was burned over in the Dolan Fire. Flows of water, mud, and debris quickly overwhelmed the small culvert that had been placed beneath Highway One to allow the normal creek flow to pass under the highway. The fill that had been placed in the drainage to support the highway was eroded rapidly, leading to 150 feet of road collapse and complete closure of the Big Sur Highway (see figure 1.1). The road, fill, and drainage beneath the road were completely rebuilt in three months at a cost of $11.5 million.

Uplifted Marine Terraces

The coastline of California has many examples of elevated marine terraces, which are characteristic features of collisional tectonic coasts where uplift has taken place. These flat benches, which typically resemble a massive flight of stairs, are usually less than a mile in width and commonly rise to elevations of 500 to 600 feet above sea level or higher.

Each terrace consists of a nearly horizontal or gently seaward-sloping wave-eroded platform backed by a steep or degraded relict sea cliff along its inland edge. The terraces were formed by wave erosion in the surf zone in the geological past, in essentially the same way that the rocky intertidal platform that is visible during low tides is being eroded by breaking waves in the surf zone today.

The number of terraces exposed along the coast of California ranges greatly: A single bench underlies the north coast town of Mendocino and the University of California, Santa Barbara campus on the central coast. There are as many as 5 benches along the north coast of Santa Cruz County and up to 13 on the Palos Verdes peninsula of Los Angeles County that culminate at an elevation of 1,480 feet. Nowhere along the entire coast of the United States is a more complete sequence of terraces preserved.

Low bluffs or cliffs that are cut into these uplifted marine terraces, which typically consist of sedimentary rocks such as shales, mudstones, and sandstones, characterize much of the coastline of San Diego, Orange, Santa Barbara, San Luis Obispo, Santa Cruz, and San Mateo Counties. Along the Northern California coast, portions of the Sonoma, Mendocino, Humboldt, and Del Norte County coasts also expose uplifted marine terraces. The distribution of these nearly horizontal marine terraces has enabled California's intensive coastal development to take place. Ease of access and construction attracted developers, and the past 75 years saw the subdivision of thousands of acres of marine terraces in Southern California.

The weak sedimentary rocks that lent themselves to wave erosion, resulting in the formation of these nearly flat terraces in the geological past, are the same materials exposed in the coastal cliffs today. These rocks are very susceptible to erosion by waves during very high tides, as well as by rainfall, runoff, and occasionally rockfalls or landslides, and thus the cliffs continue to retreat landward. Depending upon the material making up the cliff or bluff and its inherent weaknesses, erosion or retreat may proceed as small fragments or rocks fail and drop to the shoreline below or as much larger masses fail (see figure 8.5). While it's not common, coastal cliffs and bluffs can also fail more catastrophically as large landslides or slumps that can carry roads, homes, or other structures to the shoreline (figure 9.7 and see figure 8.8).

While investigations of coastal erosion are usually reported as average annual retreat rates in centimeters, inches, or feet per year, it's important to understand that these are averages measured from historical maps or

FIGURE 9.7. Coastal cliff failure as a large landslide at Torrey Pines. © *2006 Kenneth and Gabrielle Adelman, California Coastal Records Project, www.californiacoastline.org.*

aerial photographs. In reality, nearly all coastal cliff or bluff retreat is an episodic process, where a large block or slab fails instantaneously, and these failures are then averaged over time. So while a home may be set back 20 feet from the edge of a coastal cliff that has been retreating at an average rate of a foot a year, suggesting that there may be 10 or more years of stability, it could all collapse in one major failure.

Continued global warming, with the associated melting of ice caps and rising sea level, along with continued and likely more intense wave attack, will progressively move these weak bluffs and cliffs landward over time. This has and will continue to present challenges to those communities and landowners who have developments or homes on these gradually but consistently retreating features (figure 9.8).

Beaches, Bays, and Estuaries

A little over 300 miles of California's coastline (28 percent) is of low relief and consists of sandy beaches, occasionally backed by sand dunes, or lowlands that may be below sea level. These settings include bays, estuaries, lagoons, and other coastal wetlands, many of which have been completely altered by development. Much of the eastern portion

FIGURE 9.8. Despite last-minute attempts to stabilize the highly erodible sediments making up these cliffs in Pacifica, these apartments were ultimately condemned and demolished. © *2010 Joel Avila, Hawkeye Photography.*

of the Los Angeles County shoreline falls into this category, including the broad beaches of Santa Monica and Redondo Beach, and the Ballona Creek wetlands. The coastline between the Ventura River mouth and Point Mugu in Ventura County is also one of low relief, as is most of the Monterey Bay coast. Beaches backed by bays, lagoons, and wetlands rather than cliffs or terraces characterize the area between Long Beach and Newport Beach. A number of these bays or lagoons have been converted to popular waterfront communities with boat access, such as at Alamitos Bay, Huntington Harbor, Marina del Rey, the Channel Islands, and the Ventura marinas.

Most low-relief areas along the coast are the broad floodplains and channels of the rivers and streams that originated in the coastal mountains and deposited sediments over hundreds of thousands of years. Some streams are quite small, so their channels and floodplains are narrow and the beaches at their mouths are limited in extent. Others, owing to the size of the drainages and the low relief near their mouths, have built wide floodplains. In Southern California these have been completely urbanized. The Los Angeles basin, now home to about 10 million people, was formed over time by the combined floodplains of the Los Angeles, San Gabriel, and Santa Ana Rivers. Because of channelization, dams, reservoirs, water withdrawals, sand and gravel mining, and other alterations, these rivers bear little resemblance to the original channels of 150 years ago.

The mouths of many coastal rivers or streams, however, are now bays, lagoons, or estuaries. During the low sea levels of the ice ages, California's streams crossed the exposed continental shelf, eroding deep canyons, and discharged 5 to 25 miles farther to the west. Each time an ice age ended, glaciers retreated and ice sheets melted, causing a rise in sea level, usually several hundred feet. As sea level rose, the streams migrated back across the shelf, and today, in a climatically warm or interglacial period, sea level is higher than it has been for about the last 120,000 years. The mouths of most coastal streams have been drowned or flooded by high sea level, forming lagoons or estuaries. Sediment carried by the streams has been deposited and reworked by wave action and wind into sand spits and dunes and, in some cases, altered by engineering. Humboldt Bay, Bolinas Lagoon, and Morro Bay are good examples, as are those many coastal lagoons of Southern California, such as Mugu Lagoon, Batiquitos Lagoon, Agua Hedionda Lagoon, Balboa Bay, and San Diego Bay. Other lagoon areas have been completely altered by filling and reclamation, followed by development, and no longer bear much resemblance to their original character and configuration: Marina del Rey, Alamitos Bay, Huntington Harbor, and Mission Bay are good examples.

Many of these low-lying shoreline communities were developed within a few feet of sea level without the knowledge that the level of the ocean would rise over time, and that their locations lay within a long-term pattern of sea-level rise and fall (figure 9.9). Sea level is directly connected to global climate, with high sea levels such as today's resulting from a warmer climate, which both melts land ice and also heats ocean water, causing it to expand. Cooler or glacial periods cause sea levels to drop as more ocean water is sequestered on land in ice sheets and glaciers and as the oceans cool and contract.

Sea level has risen and fallen 350 to 400 feet (100 to 125 meters) many times over the last several million years in response to global climate change (see figure 9.9). Following the end of the last ice age about 20,000 years ago, sea level rose at an average rate of about half an inch per year until about 8,000 years ago, when it essentially leveled off. Since the beginning of the Industrial Revolution roughly 180 years ago, and

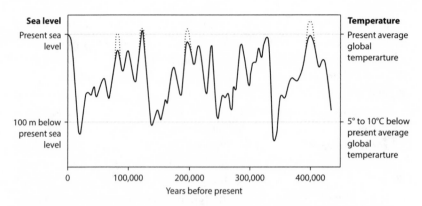

FIGURE 9.9. Rise and fall of global sea level over the past 450,000 years. Dotted lines are higher levels recorded in several geographic areas.

the burning of first coal, then oil and gas in increasing amounts, the carbon dioxide content of the atmosphere has increased about 50 percent. Global temperatures have continued to rise, and sea level has followed.

In contrast to either wave attack at times of high tides or sea level elevation that leads to the erosion and retreat of coastal cliffs and bluffs, it is both the short- and long-term rises in sea level that are having great impacts on California's low-lying shoreline communities, through flooding.

In the short term—years to decades—it will be the occurrence of the highest yearly tides, including the extreme highs now known as king tides, that will continue to be most damaging to low-elevation shoreline development, whether roads, highways, parks, or homes and businesses (see figures 9.1 and 9.4). These highest tides can be further amplified if they occur during El Niño years, when sea levels along the state's coastline may be elevated by a foot or two.

Over the long term, however, beyond about midcentury, it will likely be sea-level rise that will be the most threatening and damaging to those low-lying coastal communities. Some of these areas are already being flooded at very high tides today. Based on the best available science to date on future sea-level rise, about 3 feet of additional sea level by 2100 is the mean projected scenario. Other projections, however, with higher levels of greenhouse gas emissions, could lead to as much as 6 to 10 feet of additional sea level by 2100. California has a lot of coastal cities—61 of them to be precise, particularly in Los Angeles and Orange Counties, from Imperial Beach to Newport Beach–Balboa, and the Long Beach Peninsula, Naples, and Belmont Shore—that are highly vulnerable to very high tides today and will be more so with future sea-level rise. While flooding of these areas will not be immediately catastrophic like a major earthquake, over the second half of this century and beyond, what began as occasional high tide or nuisance flooding will become permanent and damaging inundation.

Based on accurate elevation maps and projections of future sea levels, those areas that will be susceptible to future sea-level rise can be identified. The gradual flooding of any of California's coastal communities at incremental rises in sea level can be delineated by using either Climate Central's Surging Seas website (https://sealevel.climatecentral.org) or NOAA'S Sea Level Rise Viewer (https://coast.noaa.gov/digital-coast/tools/slr.html).

SOME FINAL THOUGHTS ON COASTAL HAZARDS AND RISKS

As difficult as it may be to accept, there is absolutely nothing we can do over the long term to stop the Pacific Ocean from rising and advancing towards us, short of instantaneously turning around climate change and reversing the emission of greenhouse gases. With a herculean global effort to rapidly eliminate our burning of fossil fuels and move to renewable energy sources, we can attempt to slow climate change and sea-level rise over decades, but to date, this has eluded us. We are still dependent on fossil fuels for about 80 percent of our global energy, and we have made precious little progress in changing this pattern. The more coal, oil, and gas we continue to burn, the more the planet will warm and the more sea level will rise. This will lead to an increase in coastal cliff and bluff retreat and more frequent flooding, followed by permanent inundation of low-lying coastal development and infrastructure.

There are short-term solutions that we have employed for decades, primarily armoring the shoreline with seawalls or rock revetments or putting more sand on beaches. Both of these are very expensive and have their limits. They may forestall cliff or shoreline retreat for several years, a few decades, or longer, but with the acceleration of the rate of sea-level rise that we have documented, it is just a matter of time before individual shoreline developments or communities, highways or rail lines, wastewater treatment facilities or thermoelectric power plants, will be flooded, inundated, or damaged and destroyed. Coastal communities and state agencies are beginning to determine their most vulnerable facilities, infrastructure, and developments, with the next step being the development of realistic adaptation plans well in advance of the time when either the bluff edge is right out the back door or seawater is filling the living room. All of California's coastal cities and counties as well as state agencies are now struggling with how to plan for and adapt to this inevitable future. Establishing and agreeing on trigger points or thresholds when stepping back and relocating both public infrastructure and private development will be necessary is a critical first step in that process.

Where Do We Go from Here?

California is a study in contrasts, with the myth of paradise contrasting with the menace of natural disasters; it's the way it's always been and always will be. There won't just be the one "big one"; there will be many more big earthquakes. With the state sitting helplessly astride an active boundary between two massive tectonic plates, there is absolutely nothing that we can do to alter this geologic setting. From our human perspective, however, there are really no disasters without people, and California has an ample supply of both. And we have the power, individually and collectively, to live more safely with these hazards. But it will never be easy or straightforward, as there will always be many competing interests on how to use the state's land and resources. Political will is necessary to make any important changes in how we respond to each natural disaster and whether or not or how we rebuild in precarious locations.

California's geologic hazards are almost everywhere in the state, and scarcely a year goes by without a significant damaging natural disaster, whether a flood, fire, or earthquake. Yet people have continued to move to California in ever-increasing numbers since gold was first discovered in 1848. The promise of digging wealth out of the ground or providing supplies or services to those who did, as well as what was perceived as a near-perfect climate for people and agriculture, led to the lore of the Golden State becoming a dream; whether true or not seems to be an individual perception.

My mother's family arrived in California exactly a century ago in 1923 from Port Angeles, Washington. Black gold, or oil, was discovered at Signal Hill in Long Beach, and that promised new business opportunities for the family. My father came to Southern California from Illinois with his parents in 1914 at the urging of relatives who had relocated here. Neither family ever left, nor did their children, or their children's children, which would include me. California agreed with all of them, and they arrived in Southern California at a time when development was booming, land and homes were relatively inexpensive, and opportunities were vast. This seemed to be the case for most new immigrants, whose friends, families, or relatives often followed them to California after hearing stories of no winters and ample year-round fresh produce.

In 1920, about the time that both of my parents arrived, the state had just 3.6 million people, but by 1941, immigration and births had doubled the state's population to 7.2 million. It nearly doubled again to 14.3 million by 1957, and then doubled again by 1988. Today, California has 39 million residents within its borders. Los Angeles County in those early years was the epicenter of California's population growth, expanding from 936,000 in 1920 to 7,477,000 by 1980. During that era of rapid expansion, a major earthquake damaged Long Beach in 1933, and serious Southern California flooding occurred in 1938, 1939, 1963 (from a dam failure), and 1969. Losses were significant in each of these events, but it didn't dim the attraction of the state, nor did it slow the growth and suburban expansion.

As growth and urban sprawl have continued throughout California, the patterns and practices have been similar. Those sites that were the easiest to develop and build on were exploited first, although these weren't necessarily the safest. Flat floodplains adjacent to rivers and streams provided level building sites and fertile soil, but early residents discovered soon enough that these areas were where rivers spread out during times of high rainfall and runoff. The now urbanized alluvial fans at the base of the mountains surrounding the Los Angeles basin were built by debris and mudflows from the steep surrounding hills during periods of torrential rainfall, now known as atmospheric rivers. With the flatter land used up around urban areas in Southern California and the greater San Francisco Bay area, developers and builders moved up into the steeper terrain, often brush or tree covered, which was both landslide and wildfire prone. Many of these sites had great views, and the hazards and risks weren't obvious to immigrants from beyond California. There was money to be made, and the incentives for developers to build and real estate agents to sell properties and homes overshadowed any concern with geologic hazards and risks at that time.

Californians seem to have short "disaster memories"; after each catastrophe, our priority is to quickly repair the damaged roads and utilities, repair or rebuild our homes, and get back to "normal," often without learning or retaining anything from what just transpired. Because of an understandable connection to their homes and neighborhoods, people have rarely thought to consider that we need to adapt to natural hazards rather than trying to subdue or control them. In a battle between you and the Earth, bet on the Earth—Mother Nature always bats last. But instead, whether along a Florida shoreline just ravaged by a hurricane, or at a forested community in California recently devastated by a fire, the focus is, more often than not, how to simplify or expedite the building permit process and collect insurance so those who have lost homes can rebuild, often in the same areas and on the same sites. And most politicians historically have been sensitive to these issues. They have tended to respond with quick fixes to help people repair their lives and property rather than to look, longer term, at more permanent solutions to remove people and development away from hazardous settings. This will never be easy, but each new climate or geologic disaster will likely bring greater damages and higher fatalities. This trend will continue unless there are community- or state-level land use decisions, as unpopular as they may be, to develop responses to progressively reduce the inherent risks we face as a state.

This issue is at the intersection today of private property rights, the role and power of local governments, the challenges facing the insurance industry, and the real estate profession, all with separate interests, responsibilities, and priorities. Even though homes or entire neighborhoods may have been damaged or destroyed, many if not most families want to return to their properties and what's left of their homes and rebuild their lives, holding on to the memories and dreams. City and county planning departments are caught between the responsibility of serving and assisting impacted residents and also the obligation to protect them from hazards that the public may not fully understand or appreciate.

To be perfectly clear, insurance companies are in business to make money, and for many years the rates they charged, particularly in the most hazardous areas, didn't reflect the risks and actual costs of home replacement. In addition, many policies have fine print exclusions for certain hazards, floods and landslides for example, which are rarely read. FEMA, the Federal Emergency Management Agency, typically picked up the additional costs when an area had been declared a disaster area. So rebuilding was being subsidized by the rest of us. With the vast losses in the huge wildland fires of recent years in California, the insurance companies are being forced to reevaluate their policies and practices. Without insurance, home loans or mortgages will be far more difficult if not impossible to obtain, which may begin to deter rebuilding in the riskiest areas. Thus, economics may become a more effective solution to building or rebuilding in hazardous areas in the future. As a recent response to these risks, State Farm General Insurance Company—America's largest home insurer by premium volume—made the decision, in May 2023, to halt the sale of new home insurance policies in California, citing wildfire risk and inflation of construction costs. Allstate, the state's fourth-largest insurer as of 2021, stopped selling home insurance in California in 2022.

Home insurance is an essential part of purchasing a home. Mortgage companies normally require proof of insurance before approving a loan, in order to protect their investment in the home. Without insurance, buyers in most cases would need to make an all-cash purchase.

Real estate agents, as in most professions, come with a wide range of care, communication skills, and caution. While agents now have full disclosure requirements when homes are being advertised or sold, it is typically up to the buyer to wade through the multitude of forms and conditions in fine print that require signatures, to satisfy themselves that the area is not subject to any natural disasters. And it's probably fair to say that a decade after a major fire, flood, or landslide, this isn't the first thing on a potential buyer's mind in the excitement about their home purchase and in what they perceive as paradise or something close to it.

We have lived for decades with the comforting assumption that there are engineering fixes for virtually any geologic hazard—levees along riverbanks, seawalls and revetments on beaches, fire sprinklers inside our homes (why inside instead of on the roofs in wooded areas?), and better-engineered structures to withstand earthquakes. But everything we build today has its limits—all protection ends somewhere. A near-disastrous recent example was the failure of the Oroville Dam spillway in 2017 due to a very-high-discharge event, which led to the evacuation of 188,000 downstream residents. We simply cannot afford to provide protection for all hazards or extreme events. Engineers build levees, for example, to withstand some reasonable design flood (often the projected 100-year event), but as history shows, a larger flood can and will occur, and levees along the Sacramento River are frequently overtopped with widespread flooding. And the more protection that is built, the more development occurs in the now-protected area with the assumption that this location is now safe.

As the climate has changed, the risks posed by various climate-related processes have also changed. Observations and climate models are consistent in documenting that warm periods are getting hotter and lasting longer. As vegetation is drying out, fire hazards are increasing, with wildland fires becoming more frequent and burning larger areas. A warmer ocean means higher evaporation rates. In response, precipitation is becoming more concentrated in the winter months, leading to more flooding, landslides, and debris flows, particularly in areas recently ravaged by fires. Risks have increased in areas not previously thought to be hazardous.

Anthropogenic climate change has altered what "normal" may have been in the past (precipitation intensity, frequency, and duration, and wave heights and sea-level rise, for example), so design events used in engineering, such as the 100-year flood or rainstorm, are not the same as they were a century ago. As a result, infrastructure that was planned by using historic conditions or assumptions may no longer be appropriate or adequate. And to be clear, we are not at some stable condition or "new normal"; rather, we are in a trend, with all natural hazard processes very likely to continue changing.

There are many lessons to be learned from our history, observations, and experiences with natural disasters as we look to the future and ask where do we go from here? As stated earlier, there are no natural disasters without people. The Earth has been in existence for about 4.5 billion years, and tectonic plates have been shifting around throughout most of this long period, accompanied by earthquakes, large and small, as well as tsunamis along shorelines. Volcanoes have been intermittently erupting, sea levels have risen and fallen, and floods have repeatedly washed over vast areas. The original inhabitants of California recognized these processes and didn't build permanent settlements in hazardous areas. As small settlements developed into villages and eventually became towns, infrastructure was built and small towns became permanent cities, which is where we find ourselves today.

Where we have already built in hazardous locations, our choices are somewhat limited. Historically we have chosen to fight nature. Sometimes we have survived, at least for a while, but as earthquakes, floods, and fires over the past century or so have shown, natural processes will continue to occur as they always have, particularly in an area as geologically and hydrologically active as California. Climate change has added another variable to the forces of nature, and one we are going to be dealing with for decades and likely centuries to come.

In 1969, over 50 years ago, Ian McHarg, a Scottish landscape architect and environmental planner, wrote a book—*Design with Nature*—that is as timely now as it was then. By "design with nature" McHarg meant that the way we occupy and modify the Earth is best when it is planned and designed with careful consideration of the natural processes and character of the landscape. His essential message was that no human action, be it building a highway or condominium complex or laying out a city or park, should proceed without a study of its suitability for the topography, vegetation, waterways, wildlife, and other natural features and processes of a region.

We find ourselves in a difficult and challenging place in California today. Our towns and cities have been built, and most are still expanding and, in many cases, without much if any attempt to design with nature. As a result, we have frequently and increasingly suffered the consequences. But all is not lost. There are many safe places to live in the state, and with a little care, we can find and enjoy those places, although housing density increases are likely to be required in the future. And, for the first time in history, the state's population is actually declining. However, as history has reminded us, and the preceding chapters have hopefully illuminated, there are many hazardous or high-risk locations where we need to make intelligent decisions on how we are going to either relocate or replan neighborhoods or adapt to the existing hazards and the changes that are coming with a warmer planet.

We need to have hope, and I think one way to achieve this is through what I call sober optimism. Yes, we do have damaging earthquakes periodically in California, but our building codes have improved very significantly over the past 50 years due to the lessons learned from past earthquakes. Newer buildings are much better able to resist seismic shaking than those built decades ago. And it is not earthquakes that kill people, but collapsing buildings, and we still have many older and substandard buildings to deal with.

The greatest natural disaster risk to life in California, without a doubt, is from earthquakes. For some perspective, however, over the past 100 years there have been about 360 fatalities, or on average just 3.6 deaths per year, from earthquakes in California. For the 39 million people in the state, this is an astronomically low risk.

For comparison, Californians suffered 3,558 automobile accident deaths in 2020, or almost 10 every day. In that same year, motorcycle accidents claimed 539 lives, 958 pedestrians lost their lives, and 133 bicycle riders died from accidents. There are far riskier things to worry about than dying in an earthquake in California. Older houses and other buildings will be damaged in large earthquakes, although there are many ways that these homes can be retrofitted and earthquake proofed.

Wildland fires have produced the most home losses over the past several decades, and with our warming climate, this will no doubt continue to be the case. While dense vegetation around homes in wooded areas can be cleared, and fireproof roofs and siding can be used in new homes, our expansion into wildlands will still present clear dangers. Most other geologic hazards—landslides on steep and unstable slopes, erosion of coastal bluffs and cliffs, and flooding of beach-level homes under high tide and storm wave conditions exacerbated by sea-level rise—will always be issues. While Mother Nature will eventually prevail, you can make conscious and informed decisions about whether to buy or build in these environments, and then weigh your chances to determine what level of risk you are willing to take.

By virtue of our young and active geological setting, California in many ways is an earthly paradise, but along with this there will always be geologic perils, most of which are now recognized and understood, if not always acknowledged. What actions can we take as a society over the long run to significantly reduce our exposure and future losses? Can the insurance and mortgage industries impact our future in a positive direction? These businesses are risk averse and will likely become more so as climate continues to intensify many of the geologic and hydrologic hazards we must learn to coexist with. Our decisions and choices need to reflect our collective values and thoughts.

Sober optimism can guide us, but *aggressive incrementalism* at local and state levels has been moving us slowly forward. There are essentially three approaches for responding to a major natural disaster: (1) build back better, (2) legislate or adopt new land use and building standards for hazardous areas, and (3) prevent reconstruction after a disaster in the most hazardous areas. There are examples of each of these that California has adopted over time at the state and local government levels.

The 1933 Long Beach earthquake damaged 120 public schools and destroyed 70 of those. While 120 people died in that shock, fatalities would have been astronomically higher if schools had been in session at the time. The Field Act was passed by the state legislature within a month of the earthquake, requiring new construction standards for schools, although local governments had 40 years to bring existing schools up to code. There is typically a short time window after a disastrous event when the images are vivid, pressure is on, and there is incentive and public pressure to pass legislation to change how, where, and whether we build.

It was clear from the impacts of the 1971 San Fernando earthquake that many older public buildings (hospitals in this case), dams, and freeway overpasses were underdesigned for the forces that occurred even during a moderately large (6.6 magnitude) earthquake. New seismic standards and retrofitting followed San Fernando and have produced safer infrastructure and buildings.

After the disastrous Love Creek landslide in the Santa Cruz Mountains in January 1982, the Santa Cruz County Board of Supervisors made a difficult decision to declare the hillside adjacent to the landslide, where additional cracks had been discovered, unsafe to occupy. Homes were condemned in order to avert a future disaster. While not popular with the residents in the area, this was the right decision to reduce risks from any future landslides.

In the city of Pacifica, as coastal bluffs continued to encroach closer to three large apartment buildings until they were literally on the edge of a near-vertical 100-foot-high cliff, the city took a difficult but wise step and posted these buildings as unsafe to occupy and then proceeded to demolish the buildings. Public officials across the state are facing decisions like this one more and more frequently as climate continues to change and as older buildings need to be retrofitted, strengthened, relocated, or removed.

There are other examples where natural hazard losses precipitated action at the state or local government level to improve or increase our collective safety and reduce potential future losses. With the effects of climate change altering the past assumptions and standards used historically in our planning and project approval processes, there is an even greater imperative to make decisions based on the best available science and risk analysis and to move beyond short-term Band-Aid approaches, as convenient and expeditious as they may be, and think and act on a longer-term basis. Our children and their children will be living in the environment we create today.

Notes

CHAPTER 2. EARTHQUAKES AND FAULTING

1. Griggs and Gilchrist 1983
2. Wikipedia 2023
3. Keller and Pinter 1996
4. Keefer 1994
5. Youd and Hoose 1978
6. Youd and Hoose 1978
7. Lawson 1908
8. Griggs 2018
9. Griggs and Gilchrist 1983, 34
10. Griggs and Gilchrist 1983, 34
11. Griggs 2018, 22–23
12. Griggs 2018, 23–24
13. US Geological Survey 1999
14. Lawson 1908
15. Griggs 2018, 24–28
16. Griggs 2018, 24–28
17. Griggs 2018, 28–31
18. US Geological Survey 1971
19. Griggs 2018, 35–42
20. Griggs 2018, 22–23
21. Griggs et al. 1992
22. Plant and Griggs 1990
23. US Geological Survey 1996

CHAPTER 3. TSUNAMIS

1. Griggs 2018, 52
2. Lander, Lockridge, and Kozuch 1993
3. Lander, Lockridge, and Kozuch 1993, 51–53
4. Lander, Lockridge, and Kozuch 1993, 56–58
5. Lander, Lockridge, and Kozuch 1993, 66–67
6. Griggs 2018, 51–53
7. Griggs 2018, 52–53
8. Griggs 2018, 54
9. Lander, Lockridge, and Kozuch 1993, 85–91
10. Lander, Lockridge, and Kozuch 1993, 85–91
11. Gonzales 2005
12. Gonzales 2005
13. Griggs 2018, 56
14. Griggs 2011
15. Griggs 2011
16. Atwater et al. 2015

CHAPTER 4. VOLCANOES AND VULCANISM

1. Miller 1989
2. Christiansen 1982
3. Clynne et al. 2012
4. Nathenson, Clynne, and Muffler 2012
5. Crandell et al. 1974
6. Crandell and Nichols 1989
7. Miller 1980
8. Nathenson et al. 2007
9. Miller et al. 1982
10. White, Ramsey, and Miller 2011
11. White, Ramsey, and Miller 2011

CHAPTER 5. EXTREME RAINFALL AND FLOODING

1. Steinbeck (1952) 2002, 6
2. Griggs 2018
3. Griggs 2018
4. Hundley and Jackson 2020
5. Griggs and Gilchrist 1983
6. Hundley and Jackson 2020
7. US Geological Survey 1956
8. US Geological Survey 1963
9. Griggs 2018, 119–26
10. Griggs and Gilchrist 1983, 202–04
11. US Geological Survey 1971
12. Ellen and Wieczorek 1988
13. Griggs 1982

14. *Independent Forensic Team Report* 2018
15. Griggs 2018, 198

CHAPTER 6. CLIMATE CHANGE AND DROUGHT

1. Ingram and Malamud-Roam 2015
2. Gelt 1997
3. Griggs 2018
4. Griggs 2018, 173–74
5. Griggs 2018, 198

CHAPTER 7. WILDLAND FIRES

1. St. John, Serna, and Rong-Gong 2018
2. Gee and Anguiano 2020
3. CAL FIRE 2019
4. Moreland 2004

CHAPTER 8. LANDSLIDES, ROCKFALLS, AND DEBRIS FLOWS

1. Plant and Griggs 1990
2. Plant and Griggs 1990, 77
3. Lawson 1908
4. Griggs 2018, 158
5. Ellen and Wieczorek 1988
6. Griggs 1982

CHAPTER 9. COASTAL STORMS, SEA-LEVEL RISE, AND SHORELINE RETREAT

1. Griggs 2010
2. Griggs, Patsch, and Savoy 2005
3. Griggs 2018
4. Griggs 2018, 80
5. Griggs 2010, 23–25

References

CHAPTER 1. INTRODUCTION TO CALIFORNIA'S NATURAL DISASTERS

Sources Consulted

Abbot, P.L. 2023. *Natural Disasters*, 12th ed. New York: McGraw Hill.

Best, D.M., and D.B. Hacker. 2021. *Earth's Natural Hazards: Understanding Natural Disasters and Catastrophes*. Dubuque, IA: Kendall Hunt.

DeVecchio, D.E., E.A. Keller, and J.J. Clague. 2014. *Natural Hazards: Earth's Processes as Hazards, Disasters and Catastrophes*. New York: Routledge. 574 pp.

Gentry, C. 1968. *The Last Days of the Late Great State of California*. New York: G.P. Putnam's Sons. 384 pp.

Griggs, G.B. 2018. *Between Paradise and Peril: The Natural Disaster History of the Monterey Bay Region*. Santa Cruz, CA: Monterey Bay Press. 198 pp.

Griggs, G.B., and J.A. Gilchrist. 1983. *Geologic Hazards, Resources, and Environmental Planning*. Belmont, CA: Wadsworth. 502 pp.

Gunderson, B.Z. 2014. *Disasters: Natural and Man-Made Catastrophes through the Centuries*. New York: Square Fish. 256 pp.

Hyndman, D., and D. Hyndman. 2014. *Natural Hazards and Disasters*. Farmington Hills, MI: Brooks-Cole/Cengage Learning. 555 pp.

Lawson, A.C. 1908. *The California Earthquake of April 18, 1906, The Report of the State Earthquake Investigation Commission*. Washington, DC: Carnegie Commission.

McWilliams, C., and L.H. Lapham. 1999. *California: The Great Exception*. Berkeley: University of California Press. 391 pp.

CHAPTER 2. EARTHQUAKES AND FAULTING

Sources Cited

Griggs, G.B. 2018. *Between Paradise and Peril: The Natural Disaster History of the Monterey Bay Region.* Santa Cruz, CA: Monterey Bay Press. 198 pp.

Griggs, G.B., and J.A. Gilchrist. 1983. *Geologic Hazards, Resources, and Environmental Planning.* Belmont, CA: Wadsworth. 502 pp.

Griggs, G.B., J.S. Marshall, N.A. Rosenbloom, and R.S. Anderson. 1992. "Ground Cracking in the Santa Cruz Mountains." In *Loma Prieta Earthquake: Engineering Geologic Perspectives,* edited by J.E. Baldwin and N. Sitar, 25–42. Association of Engineering Geologists Special Publication No. 1. https://www.aegweb.org/.

Keefer, D.K., ed. 1994. *The Loma Prieta, California, Earthquake of October 17, 1989: Strong Ground Motion and Ground Failure.* US Geological Survey Professional Paper 1551-C. pubs.usgs.gov.

Keller, E.A., and J. Pinter. 1996. *Active Tectonics: Earthquakes, Uplift, and Landscape.* Hoboken, NJ: Prentice-Hall. 338 pp.

Lawson, A.C. 1908. *The California Earthquake of April 18, 1906, The Report of the State Earthquake Investigation Commission.* Washington, DC: Carnegie Commission.

Plant, N., and G.B. Griggs. 1990. "Coastal Landslides and the Loma Prieta Earthquake." *Earth Sciences* 43: 12–17.

US Geological Survey. 1971. *The San Fernando, California, Earthquake of February 9, 1971: A Preliminary Report Published Jointly by the U.S. Geological Survey and the National Oceanic and Atmospheric Administration.* USGS Professional Paper 733. pubs.usgs.gov. 254 pp.

———. 1996. *USGS Response to an Urban Earthquake: Northridge '94.* USGS Open-File Report 96–263. pubs.usgs.gov.

———. 1999. *Timing of Paleoearthquakes on the Northern Hayward Fault: Preliminary Evidence in El Cerrito, California.* USGS Open-File Report 99–318. pubs.usgs.gov.

Wikipedia, s.v. "Lists of Earthquakes." Accessed 2023. https://en.wikipedia.org/wiki/Lists_of_earthquakes.

Youd, T.L., and S.N. Hoose. 1978. *Historic Ground Failures in Northern California Triggered by Earthquakes.* US Geological Survey Professional Paper 993. pubs.usgs.gov. 178 pp.

Sources Consulted

Bolt, B.A. 2004. *Earthquakes,* 5th ed. San Francisco: W.H. Freeman. 320 pp.

Heaton, T.H., and S.H. Hartzell. 1987. "Earthquake Hazards on the Cascadia Subduction Zone." *Science* 236: 162–68.

Perry, S., D. Cox, L. Jones, R. Bernknopf, J. Goltz, K. Hudnut, D. Mileti, D. Ponti, K. Porter, M. Reichle, H. Seligson, K. Shoaf, J. Treiman, and A. Wein. 2008. *The ShakeOut Earthquake Scenario—A Story That Southern Californians Are Writing.* US Geological Survey Circular 1324. pubs.usgs.gov. 24 pp.

Stoffer, P.W. 2005. *The San Andreas Fault in the San Francisco Bay Area, California: A Geology Fieldtrip Guidebook to Selected Stops on Public Lands.* US Geological Survey Open-File Report 2005–1127. pubs.usgs.gov. 133 pp.

————. 2008. *Where's the Hayward Fault? A Green Guide to the Fault.* US Geological Survey Open-File Report 2008–1135. pubs.usgs.gov. 96 pp.

Toppazada, T., G. Borchardt, W. Haydon, and M. Petersen. 1995. "Planning Scenario." In *Humboldt and Del Norte Counties, California for a Great Earthquake on the Cascadia Subduction Zone.* California Division of Conservation, Division of Mines and Geology Special Publication 115. 159 pp.

CHAPTER 3. TSUNAMIS

Sources Cited

Atwater, B.F., S. Musumi-Rokkaku, K. Satake, Y. Tsuji, K. Ueda, and D.K. Yamaguchi. 2015. *The Orphan Tsunami of 1700.* Seattle: University of Washington Press. 136 pp.

Gonzales, R. 2005. "California Town Still Scarred by 1964 Tsunami." NPR, November 17, 2005. https://www.npr.org/transcripts/5007860.

Griggs, G.B. 2011. "1965: The Recovery of the First Ocean Floor Evidence of Great Cascadia Subduction Zone Earthquakes." *EOS, Transactions of the American Geophysical Union* 92 (39): 325–26.

————. 2018. *Between Paradise and Peril: The Natural Disaster History of the Monterey Bay Region.* Santa Cruz, CA: Monterey Bay Press. 198 pp.

Lander, J.F., P. Lockridge, and M. Kozuch. 1993. *Tsunamis Affecting the West Coast of the United States 1806–1992.* NGDC Key to Geophysical Records Documentation No. 29. Washington, DC: US Department of Commerce. 242 pp.

Sources Consulted

Clarke, S.H., Jr., and G.A. Carver. 1992. "Late Holocene Tectonics and Paleoseismicity of the Southern Cascadia Subduction Zone, Northwestern California." *Science* 255: 188–92.

Dengler, L., and K. Moley. 1993. *Living on Shaky Ground, How to Survive Earthquakes and Tsunamis on the North Coast.* Arcata, CA: Humboldt Earthquake Education Center, Humboldt State University. 24 pp.

Toppazada, T., G. Borchardt, W. Haydon, and M. Petersen. 1995. "Planning Scenario." In *Humboldt and Del Norte Counties, California for a Great Earthquake on the Cascadia Subduction Zone.* California Division of Conservation, Division of Mines and Geology Special Publication 115. 159 pp.

CHAPTER 4. VOLCANOES AND VULCANISM

Sources Cited

Christiansen, R.L. 1982. "Volcanic Hazard Potential in the California Cascades." In *Special Publication,* edited by R.C. Martin and J.F. Davis, 41–59.

Geological Survey Special Publication 63. http://pubs.geothermal-library
.org/lib/grc/1029689.pdf.

Clynne, M. A., J. E. Robinson, M. Nathenson, and L. P. Muffler. 2012. *Volcano
Hazards Assessment for the Lassen Region, Northern California.* US Geological Survey Scientific Investigations Report 2012–5176-A. pubs.usgs.gov.
47 pp.

Crandell, D. R., D. R. Mullineaux, R. S. Sigafoos, and M. Rubin. 1974. "Chaos
Crags Eruptions and Rockfall Avalanche, Lassen Volcanic National Park,
California." *U.S. Geological Survey Journal of Research* 2 (1): 49–59.

Crandell, D. R., and D. R. Nichols. 1989. *Volcanic Hazards at Mount Shasta,
California.* US Geological Survey General Information Product. http://pubs
.er.usgs.gov/publication/70039409. 22 pp.

Miller, C. D. 1980. *Potential Hazards from Future Eruptions in the Vicinity of
Mount Shasta Volcano, Northern California.* Geological Survey Bulletin
1503. https://babel.hathitrust.org/cgi/pt?id=uc1.31822008846297&seq=12.
43 pp.

———. 1989. "Potential Hazards from Future Volcanic Eruptions in California." *US Geological Survey Bulletin* 1847. 17 pp.

Miller, C. D., D. R. Mullineaux, D. R. Crandell, and R. A. Bailey. 1982. *Potential Hazards from Future Volcanic Eruptions in the Long Valley-Mono Lake
Area, East-Central California and Southwest Nevada; a Preliminary Assessment.* US Geological Survey Circular 877. https://babel.hathitrust.org/cgi
/pt?id=mdp.39015037719823&seq=287. 10 pp.

Nathenson, M., M. A. Clynne, and L. P. Muffler. 2012. *Eruption Probabilities
for the Lassen Volcanic Center and Regional Volcanism, Northern California, and Probabilities for Large Explosive Eruptions in the Cascade Range.*
US Geological Survey Scientific Investigations Report 2012–5176-B. pubs.
usgs.gov. 23 pp.

Nathenson, M., J. M. Donnelly-Nolan, D. E. Champion, and J. B. Lowenstern.
2007. *Chronology of Postglacial Eruptive Activity and Calculation of Eruption Probabilities for Medicine Lake Volcano, Northern California.* US Geological Survey Scientific Investigations Report 2007–5174-B. pubs.usgs.gov.
10 pp.

White, M. N., D. W. Ramsey, and C. D. Miller. 2011. *Database for Potential
Hazards from Future Volcanic Eruptions in California.* US Geological Survey Data Series 661. pubs.usgs.gov.

Sources Consulted

Donnelly-Nolan, J. M., M. Nathenson, D. E. Champion, D. W. Ramsey, J. B.
Lowenstern, and J. W. Ewert. 2007. *Volcano Hazards Assessment for Medi-
cine Lake Volcano, Northern California.* US Geological Survey Scientific
Investigations Report 2007–5174-A. pubs.usgs.gov. 26 pp.

Mangan, M. T., J. Ball, N. J. Wood, J. L. Jones, J. Peters, N. Abdollahian, L.
Dinitz, S. Blankenheim, J. Fenton, and C. Pridmore. 2018. *California's
Exposure to Volcanic Hazards.* US Geological Survey Scientific Investiga-
tions Report 2018–5159. pubs.usgs.gov. 49 pp.

Robinson, J.E., and M.A. Clynne. 2012. *Lahar Hazard Zones for Eruption-Generated Lahars in the Lassen Volcanic Center, California.* US Geological Survey Scientific Investigations Report 2012–5176-C. pubs.usgs.gov. 13 pp.

CHAPTER 5. EXTREME RAINFALL AND FLOODING

Sources Cited

Ellen, S.D., and G.F. Wieczorek, eds. 1988. *Landslides, Floods, and Marine Effects of the Storm of January 3–5, 1982, in the San Francisco Bay Region, California.* US Geological Survey Professional Paper 1434. Washington, DC: US Government Printing Office. 309 pp.

Griggs, G.B. 1982. "Flooding and Slope Failure during the January 1982 Storms, Santa Cruz County, California." *California Geology* 35: 158–63.

———. 2018. *Between Paradise and Peril: The Natural Disaster History of the Monterey Bay Region.* Santa Cruz, CA: Monterey Bay Press. 198 pp.

Griggs, G.B., and J.A. Gilchrist. 1983. *Geologic Hazards, Resources, and Environmental Planning.* Belmont, CA: Wadsworth. 502 pp.

Hundley, N., and D.C. Jackson. 2020. *Heavy Ground: William Mulholland and the St. Francis Dam Disaster.* Reno: University of Nevada Press. 440 pp.

Independent Forensic Team Report: Oroville Dam Spillway Incident. January 5, 2018. Association of State Dam Safety Officials. chrome-extension://efaidnbmnnnibpcajpcglclefindmkaj/https://damsafety.org/sites/default/files/files/Independent%20Forensic%20Team%20Report%20Final%2001–05–18.pdf.

Steinbeck, J. (1952) 2002. *East of Eden.* New York: Penguin.

US Geological Survey. 1956. *Floods of December 1955–January 1956 in Far Western States: Peak Discharges.* US Geological Survey Circular 380. pubs.usgs.gov. 15 pp.

———. 1963. *Floods of December 1955–January 1956 in the Far Western States.* US Geological Survey Water Supply Paper 1650. pubs.usgs.gov. 157 pp.

———. 1971. *Floods of December 1964 and December 1965 in the Far Western States.* US Geological Survey Water Supply Paper 1866. pubs.usgs.gov. 265 pp.

Sources Consulted

Hunter, H.M. 2023. *Flooding in the Golden State: California's Battle against Flood Risk.* Independently published. 32 pp.

Kelley, R. 1988. *Battling the Inland Sea: Floods, Public Policy and the Sacramento Valley.* Oakland: University of California Press. 420 pp.

Laine, D., and J. Graff. 2015. *Too Much Water: Stories of Flooding in California.* Scotts Valley, CA: CreateSpace. 182 pp.

Mount, J.F. 1995. *California Rivers and Streams: The Conflict between Fluvial Process and Land Use.* Oakland: University of California Press. 376 pp.

Teets, B., and S. Young. 1986. *Rivers of Fear: The Great California Flood of 1986.* Hammond, IN: C.R. Publications. 128 pp.

CHAPTER 6. CLIMATE CHANGE AND DROUGHT

Sources Cited

Gelt, J. 1997. *Sharing Colorado River Water: History, Public Policy and the Colorado River Compact.* Tucson: Water Resources Research Center, University of Arizona.

Griggs, G.B. 2018. *Between Paradise and Peril: The Natural Disaster History of the Monterey Bay Region.* Santa Cruz, CA: Monterey Bay Press. 198 pp.

Ingram, B.L., and F. Malamud-Roam. 2015. *The West without Water: What Past Floods, Droughts, and Other Climatic Clues Tell Us about Tomorrow.* Oakland: University of California Press.

Sources Consulted

Arax, M. 2020. *The Dreamt Land: Chasing Water and Dust across California.* New York: Vintage. 576 pp.

Hundley, N. 2001. *The Great Thirst: Californians and Water—A History.* Oakland: University of California Press. 800 pp.

Kahrl, F., and R. Roland-Holst. 2012. *Climate Change in California: Risk and Response.* Oakland: University of California Press. 154 pp.

Kroll, D. 2012. *The Perfect Flood: Devastation, Courage and the Heroic Rescue Efforts of the U.S. Coast Guard Helicopter 1305.* Ashland, OR: Hellgate Press.

Lassiter, A., ed. 2015. *Sustainable Water: Challenges and Solutions from California.* Oakland: University of California Press. 408 pp.

Mount, J., D. Swain, and P. Ullrich. 2019. *Climate Change and California's Water.* Public Policy Institute of California. https://www.ppic.org/publication/climate-change-and-californias-water/.

National Oceanic and Atmospheric Administration/National Integrated Drought Information System. "California." Accessed 2023. drought.gov/states/California.

Reisner, M. 1993. *Cadillac Desert: The American West and Its Disappearing Water.* New York: Penguin.

Standiford, L. 2016. *Water to the Angels: William Mulholland, His Monumental Aqueduct, and the Rise of Los Angeles.* New York: Ecco. 336 pp.

CHAPTER 7. WILDLAND FIRES

Sources Cited

CAL FIRE. January 24, 2019. "CAL FIRE Investigators Determine the Cause of the Tubbs Fire." CAL FIRE News Release.

Gee, A., and D. Anguiano. 2020. *Fire in Paradise: An American Tragedy.* New York: W.W. Norton. 256 pp.

Moreland, J. March 3, 2004. "Firestorm Report Critical of Policies, Logistics." *NC Times.*

St. John, P., J. Serna, and L. Rong-Gong II. December 30, 2018. "Here's How Paradise Ignored Warnings and Became a Deathtrap." *Los Angeles Times.*

Sources Consulted

Carle, D. 2021. *Introduction to Fire in California,* 2nd ed. Oakland: University of California Press. 248 pp.

Ferguson, G. 2017. *Land on Fire: The New Reality of Wildfire in the West.* Portland, OR: Timber Press. 212 pp.

Johnson, L. 2021. *Paradise: One Town's Struggle to Survive an American Wildfire.* New York: Crown. 432 pp.

CHAPTER 8. LANDSLIDES, ROCKFALLS, AND DEBRIS FLOWS

Sources Cited

Ellen, S.D., and G.F. Wieczorek, eds. 1988. *Landslides, Floods, and Marine Effects of the Storm of January 3–5, 1982, in the San Francisco Bay Region, California.* US Geological Survey Professional Paper 1434. Washington, DC: US Government Printing Office. 309 pp.

Griggs, G. 1982. "Flooding and Slope Failure during the January 1982 Storms, Santa Cruz County, California." *California Geology* 35: 158–63.

———. 2018. *Between Paradise and Peril: The Natural Disaster History of the Monterey Bay Region.* Santa Cruz, CA: Monterey Bay Press. 198 pp.

Lawson, A.C. 1908. *The California Earthquake of April 18, 1906, The Report of the State Earthquake Investigation Commission.* Washington, DC: Carnegie Commission.

Plant, N., and G.B. Griggs. 1990. "Coastal Landslides Caused by the October 17, 1989, Earthquake, Santa Cruz County, California." *California Geology* 43: 75–84.

Sources Consulted

Campbell, R.H. 1975. *Soil Slips, Debris Flows, and Rainstorms in the Santa Monica Mountains and Vicinity, Southern California.* US Geological Survey Professional Paper 851. pubs.usgs.gov.

Davies, T., and J.F. Shroder, eds. 2015. *Landslide Hazards, Risks and Disasters.* Boston: Elsevier. 473 pp.

Griggs, G.B., and J.A. Gilchrist. 1983. *Geologic Hazards, Resources, and Environmental Planning.* Belmont, CA: Wadsworth. 502 pp.

US Geological Survey. 2004. *Landslide Types and Processes.* Fact Sheet 2004–3072. https://pubs.usgs.gov/fs/2004/3072/.

Werner, E.D., and H.P. Friedman, eds. 2010. *Landslides: Causes, Types and Effects.* New York: Nova Science Publishers. 404 pp.

CHAPTER 9. COASTAL STORMS, SEA-LEVEL RISE, AND SHORELINE RETREAT

Sources Cited

Griggs, G.B. 2010. *Introduction to California's Beaches and Coast.* Oakland: University of California Press. 311 pp.

———. 2018. *Between Paradise and Peril: The Natural Disaster History of the Monterey Bay Region.* Santa Cruz, CA: Monterey Bay Press. 198 pp.

Griggs, G., K. Patsch, and L. Savoy. 2005. *Living with the Changing California Coast.* Oakland: University of California Press. 540 pp.

Sources Consulted

Bascom, W. 1980. *Waves and Beaches.* New York: Anchor Press-Doubleday. 366 pp.

Griggs, G.B. 1994. "California's Coastal Hazards." *Journal of Coastal Research,* Special Issue No. 12: 1–15.

———. 2017. *Coasts in Crisis: A Global Challenge.* Oakland: University of California Press. 343 pp.

Griggs, G., D. Cayan, C. Tebaldi, H.A. Fricker, J. Arvai, R. DeConto, R.E. Kopp, and E.A. Whiteman (California Ocean Science Protection Council Advisory Team Working Group). 2017. *Rising Seas in California: An Update on Sea-Level Rise Science.* Sacramento: California Ocean Science Trust. 71 pp.

Hable, J.S., and G.A. Armstrong. 1977. *Assessment and Atlas of Shoreline Erosion along the California Coast.* Sacramento: California Department of Boating and Waterways. 276 pp.

Hampton, M.A., and G.B. Griggs. 2004. *Formation, Evolution, and Stability of Coastal Cliffs: Status and Trends.* US Geological Survey Professional Paper 1693. pubs.usgs.gov. 123 pp.

Keller, E.A. 2010. *Environmental Geology,* 9th ed. Hoboken, NJ: Prentice-Hall. 624 pp.

Steinhardt, K., and G. Griggs. 2017. *The Edge: The Pressured Past and Precarious Future of California's Coast.* Fresno, CA: Craven Street Books. 299 pp.

Index

Page numbers in bold refer to a figure or table.